Breaking Frame

Breaking Frame

Technology and the Visual Arts in the Nineteenth Century

Julie Wosk

 RUTGERS UNIVERSITY PRESS
New Brunswick, New Jersey

Copyright © 1992 by Julie Wosk
All rights reserved
Manufactured in the United States of America
Design by Julie Burris, index by Nicholas Humez

Library of Congress Cataloging-in-Publication Data
Wosk, Julie.
    Breaking frame : technology and the visual arts in the nineteenth century / Julie Wosk.
      p. cm.
    Includes bibliographical references and index.
    ISBN 0-8135-1924-1 (cloth)—ISBN 0-8135-1925-X (pbk.)
    1. Art and technology—History—19th century. I. Title.
N72.T4W67 1991
701'.05—dc20   92-1623
                                                                         CIP

British Cataloging-in-Publication information available

TO MY PARENTS

Contents

	Illustrations	ix
	Preface	xi
	Introduction	1
1.	The Traumas of Transport in Nineteenth-Century Art	30
2.	Art, Technology, and the Human Image	67
3.	Technology and the Design Debate	105
4.	The Anxiety of Imitation: Electrometallurgy and the Imitative Arts	126
5.	The Struggle for Legitimacy: Cast Iron	145
6.	Classicizing the Machine: Ornamented Steam Engine Frames and the Search for an Industrial Style	178
	Afterword: Into the Twentieth Century	211
	Notes	223
	Index	251

Illustrations

1. Charles Sheeler, *Classic Landscape* — 2
2. Hugh Hughes, *The Pleasures of the Rail-Road* — 5
3. Breaking frame — 6
4. William Williams, *An Afternoon View of Coalbrookdale* — 9
5. Philip James de Loutherbourg, *Coalbrookdale by Night* — 11
6. Samuel Colman, *Storm King on the Hudson* — 15
7. "Explosion of the Boiler on Board the 'St. John'" — 16
8. "Danger Signal on the Erie Railway" — 18
9. Honoré Daumier, "Le Choix du Wagon" — 32
10. George Inness, *The Lackawanna Valley* — 36
11. William Heath, *The March of Intellect* — 39
12. Gurney's steam coach — 40
13. John Leech, *Hyde Park As It Will Be* — 44
14. Robert Seymour, *Going It by Steam* — 45
15. *The Flight of Intellect; Portrait of Mr Golightly* — 50
16. *Elopement Extraordinary* — 51
17. "The Wreck of a Moody and Sankey Excursion Train" — 53
18. Honoré Daumier, "Impressions et Compressions de Voyage" — 58
19. Honoré Daumier, "Conducteur!" — 59
20. Honoré Daumier, *The Third-Class Carriage* — 63
21. Abraham Solomon, *The Return* — 65
22. Wilson Lowry after George Robertson, *The Inside of a Smelting House at Broseley, Shropshire* — 68
23. Count de Dunin, "Expanding Figure of a Man" — 70
24. James Nasmyth, *Steam Hammer in Full Work* — 72
25. Corliss engine — 74

26. Charles Graham, "Making Bessemer Steel at Pittsburgh" 75
27. "The Great Gas Meter" 76
28. John Ferguson Weir, *The Gun Foundry* 77
29. Joseph Wright of Derby, *An Iron Forge Viewed from Without* 78
30. Robert Seymour, Plate 1 of *Locomotion* 80
31. Pierre and Henri-Louis Jacquet-Droz, Lady Musician 83
32. Robert Seymour, Plate 2 of *Locomotion* 91
33. Robert Seymour, *Living Made Easy* 93
34. Robert Seymour, *The March of Intellect, The Mechanical* 95
35. Robert Seymour, *Shaving by Steam* 96
36. *The Shaving Machine sold by D. Merry and Sons* 97
37. Cham (Amédée de Noé), Man in Turnstile 99
38. "The Lathes" 102
39. Lewis Hine, *Powerhouse Mechanic* 103
40. Middletown Silver Plate Company, "The Barge of Venus" 122
41. Elkington, Mason & Co., "The Triumph of Science and the Industrial Arts" 127
42. Reed & Barton, "The Progress Vase" 128
43. Meriden Britannia Co., Plated wine cooler and punch bowl 135
44. Christopher Dresser, Electroplated teapot 143
45. Carron Co., Cast-iron neoclassical stove 148
46. New York Wire Railing Company, Cast-iron hall chairs 150
47. Coalbrookdale Co., Cast-iron deerhound hall table 151
48. Joseph Nash, *Opening the Great Exhibition* 153
49. J. P. Gaynor, Haughwout Building 154
50. Cast-iron column capitals 165
51. Coalbrookdale Co., Cast-iron hall stand 173
52. Coalbrookdale Co., Cast-iron garden bench 175
53. Boulton and Watt rotative beam engine, "Lap Engine" 180
54. Engine for the steamship *Tiger* 183
55. John Prince, Double-barreled air pump 185
56. High-pressure steam engine 187
57. Compound beam engine 188
58. *Yankee Girl*, high-pressure steam engine 189
59. Thomas Bury, *Exterior of Euston Station* 192
60. Contrasting Greek columns, Parthenon and marine engine 196
61. Atwater sewing machine 199
62. Morton Livingston Schamberg, *Telephone* 213
63. Charles Sheeler, *Rolling Power* 218
64. *Machine Art* (exhibition catalog cover) 219
65. Charles Moore, Piazza d'Italia fountain 220

Preface

I first became interested in writing this book during a visit to a London print shop, where I discovered an extraordinary print of Robert Seymour's *Going It by Steam*—an engraving dating from 1829 in which startled human bodies have been shot up into the air by a steam explosion. It occurred to me that this print was a telling image of the sudden, unexpected, explosive changes brought by new steam technologies during the nineteenth century. Back in New York, I found other artworks that captured a different type of explosiveness. In a Currier and Ives print celebrating the opening of the Brooklyn Bridge in 1883, orange and red fireworks shoot into the night air, capturing the euphoria and proud expectation greeting this prime nineteenth-century emblem of the nation's engineering expertise.

I set out to explore this sense of eruption, and disruption, that accompanied the arrival of new technologies in the nineteenth century as it was made manifest in some well-known, and many more lesser-known, works of art. I included illustrations from science and engineering journals, since they were often not just informational reports but also sensitive barometers of the sweeping technological changes and psychic shifts during the century.

During this same period, I also became intrigued by some anomalies in nineteenth-century design, particularly factory steam engines designed to suggest classical temples, complete with Doric columns and entablatures. I had spent two summers as an NEH fellow in twentieth-century art and architecture at Columbia and Princeton universities, where I had been thinking about various versions of classicism in twentieth-century modernist and postmodern architecture and design, and it occurred to me that those nineteenth-century factory and steamship engines as well as railroad stations with classical facades were an early pairing of classical imagery with technology, though the nineteenth-century classical ornament would change to more abstracted, classicized geometries in twentieth-century machine age design.

It seemed to me, also, that those ornamented steam engines, so ridiculed by twentieth-century modernists, had a social dimension: they suggested early efforts by manufacturers and factory owners not only to present the engines as impressive and dignified but also to dissociate them from the steam explosions that were widely reported in the nineteenth century. The classical design imagery suggested, too, that the new technologies were not just ephemeral phenomena but enduring, stately, and, most important, safe additions to the social milieu.

As I thought about these issues, I became interested in exploring the new technologies that had helped transform the world of art and design. I chose electroplating and decorative cast iron as particularly interesting, since each had received little critical attention. Each, also, had involved efforts by manufacturers to legitimize them as new technologies—technologies equal, if not superior, to original works of decorative art. Again, they were both paradigms and anomalies. They were sources of pain to a number of nineteenth-century art critics, but widely accepted by a middle-class market; they were derided by critics as vulgar and flimsy, but were successfully marketed as symbols of status and style. And as factory-made reproductions or imitations of the arts in the age of steam, they were particularly interesting as a type of antecedent to late twentieth-century, postmodernist concerns with electronic simulacra and reproductions.

This book is by no means intended to be encyclopedic; there still remain much material and varied critical perspectives for others to explore. I have focused on some of the significant aesthetic, social, and psychological factors shaping the response to new technologies by nineteenth-century artists, manufacturers, and designers. Of the many new technologies that transformed the decorative arts and generated critical controversy, I chose only a few that had received little attention and confined my focus to England and the United States, though some seminal examples by French artists help illuminate the central themes of this book.

ENTERING new territory, often with few monographs or studies to guide me, I was particularly grateful to the individuals and institutions that offered me aid and support, including grants from the National Endowment for the Humanities, the Maritime College Foundation, and United University Professions of the State University of New York. I am also particularly indebted to the following individuals for their guidance and suggestions: my editor, Dr. Karen Reeds, of Rutgers University Press; David de Haan, head of collections at the Ironbridge Gorge Museum Trust, Coalbrookdale, Ironbridge, England; the staffs of the Ironbridge Library, the Hagley Library in Wilmington, Delaware, and the British Library and Science Museum Library in London; and Robert Vogel, formerly curator of engineering at the Smithsonian Institution in Washington, D.C., who read a chapter of of my manuscript. Finally, I owe special thanks to my family and friends, including my parents Joseph and Goldie Wosk, my sisters Paula Bhimani and Toby Costas and their families, Philip Cohen, Rosalind Freundlich, Bernard Heilweil, and Marlene Phillips, and to my colleagues at State University of New York, Maritime College, including Dr. William R. Porter and Dr. Joel Belson, all of whom offered encouragement and support during my many years writing this book.

Breaking Frame

Introduction

In *Battle of Lights, Coney Island* (1914), American artist Joseph Stella captured the frenzied, explosive excitement of a new technological era. Inspired by the Italian Futurists and their heady infatuation with machines, Stella fractured his own images of New York's famed amusement park into dazzling fragments of color and light. Amidst the swirls of whirling rides, segments of girders and roller coasters, flashing signs, and rays of yellow and green light beaming out at night, Stella brilliantly summed up the early twentieth century's feverish infatuation with technology and fast-moving machines. For the Futurists, the aesthetic of explosive divisionism, of fracturing and splintering, mirrored their joyous celebration of the modern age.

America's Precisionist painters starting in the 1920s shared the Futurists' fascination with modern technology, but focused not on speeding machines but machines at rest. Charles Sheeler, one of the country's leading Precisionists, evoked a sense of stillness in his clarified and classically elegant vision of industry which saw dignity, beauty, and stature in the classicized geometries of machines.[1]

Redefining traditional landscape art, Sheeler's archetypal painting of twentieth-century industrial America, *Classic Landscape* (1931, fig. 1), presented the Ford Motor Company's River Rouge plant near Detroit as a modern Parthenon with its stately columnar smokestacks and silos. This was classicism for the modern machine age, a classicism of clean, spare machine forms bathed in a purifying white light—a pristine, immaculate factory landscape set in a timeless world of stability and calm.

Sheeler's paintings both sanctioned and sanctified the world of ma-

1 Charles Sheeler, *Classic Landscape* (1931). Oil on canvas. Collection of Mr. and Mrs. Barney A. Ebsworth Foundation.

chines. His reverence for technological wonders and his belief in the inherent beauty of machines were central tenets of twentieth-century modernism—beliefs undaunted by the jibes of Europe's Dada artists who wittily satirized machine idolatry.

These idealized images of technology suggested that the machine had become a privileged object radically assimilated into the consciousness of twentieth-century life. As Sheldon Cheney wrote in *Art and the Machine* (1936) with an air of satisfaction, "We now generally accept the machine's coming as a major evolutionary advance. The wonder of the machine need no longer be stressed. The thrill felt in its harnessed power, the admiration for its marvelously precise functioning, its adaptability, its capacity for multiplying commodities, are part of a common experience."[2]

The vision of technology as comfortably implanted in the social and cultural milieu was not at all firmly established during the nineteenth century. The twentieth-century infatuation with a machine aesthetic came after a century of disruptive technological changes, including the arrival of new steam engines, industries, railroads, and factory-made imitations of decorative arts—changes that were both welcomed and feared.³

Art in nineteenth-century Britain and America, two countries undergoing dramatic industrial growth, often mirrored the traumas felt by those living in rapidly mechanizing societies. While nineteenth-century artists shared some of Joseph Stella's explosive excitement, they also captured the sense of psychic and physical splintering caused by deadly machine accidents and explosions in the age of steam.

These artists created images of people living, or dying, in a type of technological nightmare, an existential world of contingency and chance. It was the kind of experience deftly described by America's Charles Francis Adams in his study of railway accidents written in 1879, a study that was ironically intended to reassure travelers about the safety of the railroad. Capturing the traumatic experience of some train travelers, Adams wrote, "Suddenly, somehow, and somewhere . . . an obstruction is encountered, a jar, as it were, is felt, and instantly, with time for hardly an ejaculation or a thought, a multitude of human beings are hurled into eternity."[4]

Nineteenth-century artists dramatically depicted not only the shattering effects of railroad accidents but also the experiences of bewildered people negotiating the unfamiliar territory of new technologies, people experiencing the hazards and humiliations of adapting to new machines. This feeling of disorientation when encountering new experiences has been more generally described by American sociologist Erving Goffman as the experience of "breaking frame." As discussed by Goffman in *Frame Analysis* (1974), the experience of breaking frame occurs when the basic frameworks of understanding used to make sense out of events no longer apply. The subject, "expecting to take up a position in a well-framed realm," discovers that "no particular frame is immediately applicable, or the frame that he thought was applicable no longer seems to be." In this unexpected situation, "the unmanageable might occur, an occurrence which cannot be effectively ignored and to which the frame cannot be applied, with resulting bewilderment and chagrin on the part of the participants."[5]

An instance of breaking frame occurred at the formal ceremonies celebrating the opening of Britain's Liverpool and Manchester Railway in 1830. As reported in the *Mechanics' Magazine,* a procession of railroad carriages traveled on two adjacent tracks toward Manchester carrying six hundred passengers, including the Duke of Wellington, and over thirty distinguished guests. Seventeen miles from Liverpool, the engines stopped to receive a fresh supply of water and fuel, and many of the travelers alighted from the trains, "walking about congratulating each other on the truly delightful treat they were enjoying, all hearts pounding with joyous excitement."

Among the dignitaries who were walking down one line to greet the Duke of Wellington was William Huskisson, a member of Parliament. A tragedy occurred as one of the several locomotives, George Stephenson's *Rocket*, traveled rapidly down the other line. Huskisson, who found himself between two tracks and was reportedly already in a weak state of health, became "flurried" and disoriented. He "lost his presence of mind, seemingly like a man bewildered," and was described as "alarmed and agitated." Huskisson was thrown to the ground by the *Rocket*, which ran over his leg and thigh, producing fractures that caused his death in a few hours.[6]

As suggested by Huskisson's terrible disorientation and death, railroad travel introduced a new world to which traditional modes of understanding and behaving, in Goffman's terms, no longer applied. This experience of breaking frame with its attendant feelings of disorientation and dislocation became an important new theme in nineteenth-century art, illuminating the century's undercurrents of uncertainty and fear.

The sudden, shattering experience of railroad accidents was spoofed in nineteenth-century British caricatures, where exploding steam engines were shown fracturing social frameworks and wrecking human frames. In Hugh Hughes's ironically titled engraving *The Pleasures of the Rail-Road* (1831, fig. 2), the artist both mocks and mirrors public fears about steam travel, showing a violent railroad catastrophe that has thrown wide-eyed passengers high into the air. In Hughes's print, sharp radiating lines suggest the force of the explosion and extend outward to the edges of the pictorial frame.

Through their images of broken frames—psychic, social, bodily, and mechanical—artists often brilliantly articulated the fracturing that haunted the nineteenth century's surface rhetoric of progress. But artists were not just engaged in detailing the fragmentation and divisions produced by disruptive technologies: they also imaged signs of technology's social integration, signs of a cultural rapprochement with technological change.

While Hughes's *Pleasures of the Rail-Road* with its radiating lines presented a sardonic view of technological fracturing, artists after

2 Hugh Hughes, *The Pleasures of the Rail-Road. Shewing the Inconvenience of a Blow-Up* (1831). Etching. Ironbridge Gorge Museum Trust, Elton Collection.

midcentury, as discussed in chapter 1, also produced paintings, journal illustrations, and commercial images of railroad passengers in a more bounded, secure setting beneath the ornately decorated, arched ceilings of the passenger car, hardly mindful of the speeding train.

Reading, dining, conversing, they ride comfortably ensconced in their cozy and luxurious train carriages with velvet seats and wood-paneled walls. These images of railroad travel, to a certain extent, reflected actual improvements in nineteenth-century passenger comfort and safety, yet they also bore the hallmarks of an advertiser's idealized view. In effect, they represented an alternative reality—a vision of safety and security that existed along with, rather than displaced, images of fracturing railroad collisions and runaway trains.[7]

These images of fracturing and integration, rupture and reconciliation, emerged as central themes in nineteenth-century artists' views of technology. While their work often illuminated the century's contradictory experiences, the process of creating art was in itself an integrative act: it was a way of bringing a coherent vision to the in-

6   Breaking Frame

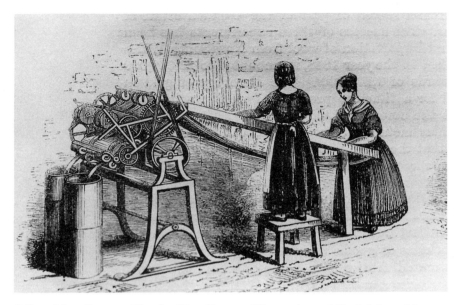

3 Breaking frame. Charles Tomlinson, *Illustrations of Useful Arts, Manufactures, and Trades* (London: Society for Promoting Christian Knowledge, 1858).

choate disruptions of contemporary encounters with technology, a way of mirroring that which was not yet fully perceived, comprehended, or understood.

The aesthetic integration of the loose, diverse strands of contemporary experience had its analogue in an actual nineteenth-century textile mechanism that was, intriguingly, called a "breaking frame" (fig. 3). The nineteenth-century textile breaking frame was used to prepare raw cotton for spinning. In the slivering process, short strands of cotton called slivers were spread flat upon a feeding board and presented to a pair of rollers on a sliver box or breaking frame, which drew them in where they were joined with other overlapped slivers into uniform, untwisted, continuous strands in preparation for spinning.

By visually clarifying the century's diverse cultural responses to technology, nineteenth-century artists were in a sense creating their own version of a breaking frame—providing a means to integrate slivers of experience in a disruptive, disjunctive industrial era. The task was not welcomed by many artists, but a small, important body of works revealed artists' abilities to weave the explosive effects of technology into a coherent framework.

The dialectic of fracturing and coherence, division and unity, rupture and reconciliation shaping nineteenth-century artists' images of technology is a major focus of this book. Entering new territory, the first two chapters look at artists' startling ability to illuminate the underlying tensions accompanying the arrival of new technologies, their ability not just to soften the century's profound sense of disorientation and trauma, but also to make it painfully visible. Through their work, nineteenth-century British and American artists offered sobering as well as comic views of people experiencing the hazards and uncertainties of riding in new steam-driven vehicles. Their art also revealed underlying fears of people themselves becoming machinelike, of people redefining their own sense of stature and scale in the presence of gigantic machines. Yet finally, artists offered a vision not just of broken frameworks but also of people adapting, though sometimes tentatively and problematically, to these new technologies which had so transformed their age.

To the British and American artists who recorded the great burst of technological innovation during the second half of the eighteenth century, all seemed a vision of heightened energy and motion, but it was a motion kept within the bounds of a harmonious frame. Erasmus Darwin, in his poem *The Botanic Garden* (1791), celebrated the new sciences, engines, and machinery transforming the age. The steam engine became a monument to both motion and order with its piston "resistless, sliding through it's iron walls" and "balanced beam, of giant-birth."[8]

John Dyer's poem *The Fleece* (1757) welcomed Britain's engagement in cotton manufacture, envisioning the emerging industrial centers of Birmingham, Manchester, and Sheffield as sites of booming construction, where neoclassical structures added a note of elegance to the new industrial era:

> . . . th'echoing hills repeat
> The stroke of ax and hammer; scaffolds rise,
> And growing edifices; heaps of stone,
> Beneath the chisel, beauteous shapes assume
> of frize and column.[9]

Dyer's poem is a clear assertion of technological legitimacy; it reveals the poet's obvious delight in the rational perfection of the machine. Viewing Lewis Paul's multiple spinning machine (patented in 1739 for use in textile mills), he praised its innovative design, which was powered by "a wheel, invisible, beneath the floor," a wheel that "to ev'ry member of th' harmonious frame / Gives necessary motion."[10]

Dyer's admiration for the "harmonious frame" that produced "necessary motion" not only gave tribute to well-designed textile machinery but also echoed eighteenth-century England's philosophical belief in a rational, harmonious universe designed as a smooth-functioning machine. In his poem, even the sheep providing wool for textile manufacture are part of this cosmic mechanism: they perform "the work of perfect reason" and "turn the wheels of nature's vast machine."[11] As Alexander Pope had insisted in his *Essay on Man* (1733–34), there existed an orderly perfection in both earthly and heavenly frames, for

> All are but parts of one stupendous whole,
> Whose body Nature is, and God the soul;
> That, chang'd thro'all, yet in all the same;
> Great in the earth, as in th' aethereal frame.[12]

The seventeenth-century Newtonian conception of a harmoniously ordered clockwork universe became a recurrent aesthetic metaphor during the eighteenth century's social, political, and technological upheavals.[13] British artists who recorded the new iron furnaces and textile mills at the end of the eighteenth century embedded the industries' intrusive arrival in an aesthetic framework of pastoral harmony and calm.

Eighteenth-century British artists were devoted to the conventional subjects of art (including landscapes, historical themes, portraits), and paintings of the new industries were relatively rare. But among the small number of paintings with industrial subject matter were views of Coalbrookdale, one of England's most important industrial sites, known for its production of pig iron and its castings of cylinders for steam engines, and as the place where Abraham Darby first used coke to smelt iron in 1709.

Coalbrookdale clearly reflected disturbing signs of industrial change. During the period 1750–1800, British travelers touring the rolling hills and woodlands of the Shropshire countryside often admired the growing iron industry but also saw signs of pollution as blast furnaces and the calcining of iron ore spread sulfurous fumes over the landscape. A visitor to Coalbrookdale in 1785 described it as "a very romantic spot" all "thickly covered with wood" but also noted a "variety of horrors" including smoke from lime kilns and "flames bursting from the furnaces" that were burning coal.[14] Another visitor compared explosive sounds of ignited gases during the iron casting to the sounds of artillery and muskets, "those horrid Engines of destruction."[15]

But while visitors to Coalbrookdale described hellish fires and polluted skies, there was no hint of the satanic in *A Morning View of Coalbrookdale*, painted in 1777 by British artist William Williams (1740–1798), whose idyllic, romanticized vision presented only a distant view of a pumping engine and the smoke from blast furnaces. The painting's industrial images are comfortably dwarfed by looming trees and rolling hills, and almost lost beneath the vast sky over the countryside. Williams's companion painting, *An Afternoon View of Coalbrookdale* (1777, fig. 4) was another panoramic view with Coal-

4 William Williams, *An Afternoon View of Coalbrookdale* (1777). Oil on canvas. Shrewsbury Borough Museum Service.

brookdale's lower furnace pool and smoke from the upper furnace seen in the distance. Several fashionably dressed figures standing high on a hill casually conversing provide an air of tranquillity as they blissfully ignore the smoky industrial site below.[16]

Williams's paintings of Coalbrookdale underplayed any signs that new industries were having a disruptive effect, but it was this same site just a few decades later that was violently disrupted by an explosion of the upper furnace in 1801, and later visitors to the ironworks continued to describe frightening explosions and "volcanic eruptions" of fire.[17] James Nasmyth, British inventor of the steam hammer, also grimly reported a disrupted rural landscape in England's coal and iron-producing Black Country in the West Midlands. Devastated by the iron industry, "once happy farmhouses" were now "ruined and deserted," and "every herbaceous object was of a ghastly gray."[18]

Reacting to such changes brought by technology, William Wordsworth in his poem *The Excursion* (1814) took a dim view of the "dizzying wheels" of his "inventive Age." While Erasmus Darwin at the end of the eighteenth century delighted in mechanical motion, Wordsworth was skeptical:

> . . . I have lived to mark
> A new and unforseen creation rise
> From out the labours of a peaceful Land
> Wielding her potent enginery to frame
> And to produce, with appetite as keen
> As that of war, which rests not night or day,
> Industrious to destroy! . . .[19]

One of the most frequently presented views of technology in nineteenth-century art is that artists often minimized the disturbing impact of new railroads and industries, softening all hints of harshness within the curves of a picturesque landscape.[20] But it is important to note that amidst these benign images, there were countervailing images revealing the disruptive, explosive presence of new technologies and people experiencing the trials and traumas of adapting to new machines.

The "potent enginery" that framed the landscape in Wordsworth's poem became a source not only of vast power but also of profound disruption. In a world of exploding steam boilers, derailing trains, and polluted skies, the image of a harmoniously designed clockwork universe was soon competing with a more discordant view.

5 Philip James de Loutherbourg, *Coalbrookdale by Night* (1801). Oil on canvas. Trustees of the Science Museum (London).

The disruptive experience of changing technologies—and also the wish to strike back—may have contributed to what became a quintessential paradigm of technological fragmentation and broken frames: the Luddite acts of violence from 1811 to 1813 in which wrathful bands of jobless British textile workers vandalized stocking- and lace-making frames, a protest against the introduction of more productive machines in workshops as well as rising prices and poor living conditions. Angry at the gig mill and the shearing frame, two labor-saving machines, the Luddites smashed a thousand frames. Continuing their outbreaks of violence, they wrecked power looms in 1826 and again in the 1840s.[21]

In their views of technology, nineteenth-century artists often captured the painful sense of fracturing and dislocation brought by technological change, and their art skillfully illuminated the subtext of dysfunction that underlay the century's vision of progress. One of the most dramatic paintings of industry's explosive presence at the very beginning of the nineteenth century was British artist Philip James de Loutherbourg's fiery *Coalbrookdale by Night* (1801, fig. 5), a powerful, emblematic image of industry's eruptive impact on the landscape. Forge furnaces, an engine house, and a casting house appear on the painting's right, but dominating the landscape are lurid red flames from the coke hearth and billowing clouds of smoke. The smoke obscures the moon and transforms the town into an industrial inferno, with the black foundry buildings starkly silhouetted against the night sky.

The fiery imagery in de Loutherbourg's painting has often been viewed in terms of late eighteenth-century aesthetic conventions of the picturesque, romantic, and sublime (an aquatint of the Coalbrookdale ironworks appeared in the artist's book *The Romantic and Picturesque Scenery of England and Wales*, 1805). As described by British political philosopher Edmund Burke, the sense of the sublime was evoked by forms and forces in nature associated with majesty—forces that created in the viewer feelings of awe, terror, and dread. During the eighteenth century the sublime was associated with images of wild, untrammeled nature, with earthquakes, mountains, fires, and storms—with the terrific and terrifying.[22] And as skillfully documented by F. D. Klingender in *Art and the Industrial Revolution*, eighteenth-century British artists' images of the sublime also often associated industry with an inferno. The fiery red sky in *Coalbrookdale by Night* echoed contemporary descriptions of industrial sites which rang with references to the demonic, heard in William Blake's memorable indictment of England's "Satanic Mills." British dramatist Charles Dibdin, observing Coalbrookdale, thought it lacked "nothing but Cerberus to give you an idea of the heathen hell."[23]

The theatricality and infernal imagery in *Coalbrookdale by Night*, however, were not just echoes of the sublime: they also reflected the artist's earlier experience as a theatrical scenery designer. Born in Strasbourg in 1740, de Loutherbourg later settled in London in 1771 and worked for David Garrick and playwright Richard Sheridan at Drury Lane from 1773 to 1785, where he developed imaginative, mechanized stage effects for re-creating volcanic eruptions and the light of moon, sun, and fire.

The artist's theatrical work revealed his fascination with both machinery and explosions. In 1781 he created a sensation by exhibiting for the first time a show of moving pictures, which he called the "Eidophusikon," presenting lamp-lit scenes on a shallow stage and using special mechanical and optical effects. The eidophusikon's scenes included "Sunset," "Moonlight," and one titled "Satan arraying his Troops on the Bank of the Fiery Lake, with the raising of the Palace of Pandemonium," a reference to the building of Pandemonium, the capital of Hell, in Milton's *Paradise Lost* (1667).[24]

The eidophusikon's re-creation of Pandemonium, with its shrieking demons and an erupting volcano, ingeniously embodied several of the central cultural themes associated with the industrial revolution: the fascination with mechanical devices, imagery of the inferno, and references to the uproar or pandemonium of factories and machines. The eidophusikon, as well as de Loutherbourg's painting of Coalbrookdale, thus were more than a mere reflection of contemporary aesthetic conventions: they were ultimately a powerful evocation of the awe and anxiety inspired by the emerging new industries.

During the nineteenth century, the experience of pandemonium extended beyond furnaces and factories. With the arrival of railroads, steamboats, and mass-produced goods, the new technologies brought an added sense of dislocation and uproar. While British artists satirized the new inventions and recorded the presence of polluted skies, American paintings of industry were rare until the 1830s, reflecting in part the countries' differing rates of technological development.[25] By the 1830s and 1840s, though, America had greatly increased its iron production and use of steam power in industry, and American artists including John Gabsby Chapman in his painting *Blast Furnace at the West Point Foundry* (1837) were recording signs of industrial change.[26]

Still, American painters tended to minimize the disruptive presence of new industries and the railroad, suggesting the country's commitment to technological progress and economic development.[27] After 1830, America's mania for railroad building was fueled by a variety of factors, including the need to improve land transportation, to bridge the country's great geographic distances, and to further national expansion and economic growth.[28]

In a cultural climate eager to promote new modes of transportation, American artists rarely satirized the more dangerous aspects of railroad or steamship travel. While British caricaturists were working within a tradition of social satire in the graphic arts established at the end of the eighteenth century, American artists had little in the way of a similar tradition or even significant markets for satire; several important American comic weeklies did not even appear until the 1870s.[29] Even when American artists did see some comic aspects of steam travel, their humor was rarely harsh. Occasional cartoons about railroads in *Harper's Weekly* and in Currier and Ives lithographs of the 1880s took a light view of passengers dashing out of a train during a five-minute refreshment break or lurching through trains in motion.[30]

In their views of industry, American artists at midcentury, including Sanford R. Gifford in *Landscape with Stottsville in the Distance* (c. 1850s), were likely to keep factory smoke comfortably far back in an idyllic tableau. Largely absent were visions of an industrial inferno.[31] More often, American painters, like the eighteenth-century British painter William Williams, tended to embed industrial imagery and railroads into a harmonious pictorial frame of tranquil rural scenery. The incorporation of the machine in the garden, as Leo Marx so deftly described, suggested the persistence of America's pastoral ideal.

Amid these images of absorption, though, the dialectic of fracturing and integration still permeated America's cultural responses to technology. The dual themes were often jointly presented or tentatively resolved through the agency of the artist, who produced an aesthetic model of integration. Particularly revealing was Ralph Waldo Emerson's view of industry and railroads as "disagreeable facts" that threatened to break up the integral wholeness of nature. His essay "The Poet" urged American poets to integrate these fragmentary technological disjunctions, to harmonize these aspects of the "artificial" and the natural world: "Readers of poetry see the factory-village, and the railway, and fancy that the poetry of the landscape is broken up by these; . . . but the poet sees them fall within the great Order not less than the bee-hive, or the spider's geometric web. Nature adopts them very fast into her vital circles, and the gliding train of cars she loves like her own."[32]

Emerson's vision of reconciliation is, to some extent, an aesthetic act, a mode of perception envisioning technology's integration. But American artists also presented an alternative view—a balanced, dialectical vision in which images of technology, rather than being comfortably absorbed or integrated into the landscape, are counterpointed by images of nature, balancing the nation's commitment to technological progress with its persisting pastoral ideal.[33]

An elegant and lucid distillation of this theme appears in Samuel Colman's painting *Storm King on the Hudson* (1866, fig. 6), where the curling smoke pouring from steam chimneys heads on a diagonal to the left while clouds above the looming mountains in New York form a diagonal to the painting's right. Linking the two poles of the painting, the small central figures seated in a boat in the middle distance create an emblematic bridge between technology and the natural world.

6 Samuel Colman, *Storm King on the Hudson* (1866). Oil on canvas. National Museum of American Art, Smithsonian Institution, Gift of John Gellatly.

Like their British counterparts, American artists also enjoyed trumpeting technology with soaring, larger-than-life images that mythologized engineering achievements. *Scientific American,* which began publication in 1845, and American weeklies including *Harper's* published wood engravings and lithographs of mammoth machines that mirrored the country's gargantuan technological aspirations. American graphic artists, again like the British, also documented the century's great industrial exhibitions including the New York Crystal Palace Exhibition of 1853 and Philadelphia's Centennial Exhibition of 1876, where proudly displayed emblems of engineering prowess satisfied the nation's hunger for icons of progress.

Yet for all their mythic images of technological gigantism and tranquillity, American artists also showed the traumatic effects of new technologies. Although American painters continued to shy away from showing the more traumatizing effects of technological change, graphic artists working for American and British weeklies including the *Illustrated London News* (starting in 1842) had an eager audience for their dramatic scenes of devastating railroad and steamship accidents. These newspaper illustrations as well as separately issued prints often presented both reportorial as well as sensationalized views of people breaking frame—people disoriented and trauma-

7 "Explosion of the Boiler on Board the 'St. John,' October 29, 1865—Scene in the Main Saloon" (sketched by John P. Newell). *Harper's Weekly*, November 18, 1865.

tized by technologies and machines. Charles Francis Adams, railroad commissioner for Massachusetts and president of the Union Pacific, blamed these sensationalized reports for inflaming the public imagination: when a railroad accident occurs, he complained, "it is heralded like a battle or an earthquake; it fills the columns of the daily press with the largest capitals and the most harrowing details," making a "deep and lasting impression on the minds of many people."[34]

Steamship boiler explosions were also reported in journals, which used melodramatic rhetoric and images to describe the catastrophes. The illustration accompanying a *Harper's Weekly* story in 1865 of an explosion on the Hudson River steamer *St. John*, in which a burst boiler issued thirty tons of scalding water, showed distraught women wading through the flood and architectural pillars bathed in billowing steam (fig. 7). The rush of steam that poured into the ship's stateroom, wrote *Harper's*, created a "heart-rending" scene in which "frightened passengers leaped barefoot into the hot water" and "poor struggling creatures lay helpless and bruised, tossing in unutterable agony."[35]

American and British artists were also sensitive to difficulties posed by the new technologies, including the sensations felt by people riding in early railroads who were adapting to faster mechanical speeds. For some, the ride produced the exhilarating feeling of flying. But railroad travel and rides in the new experimental steam carriages at the end of the 1820s also brought fears of being at the mercy of runaway machines speeding out of control. Prints showing collisions and derailments became an insistent reminder of the railroad's unswerving, often catastrophic course. Artists were also sensitive to the emotionally charged experience of brake failure, of riding in machines whose frames could not easily be halted.

Britain's first scheduled freight-carrying railroad, the Stockton and Darlington Line, had no braking mechanism on the engine, and the braking systems on early nineteenth-century railroads remained relatively primitive. Until George Westinghouse patented his air brake in 1869 and the brakes were universally adopted, American railroad engineers between 1830 and 1875 used manually powered brakes and waited tensely for the "down brakes" signal, when they were forced to apply individual brakes at the end of each railway car. If the brakes failed to act simultaneously or if the train's load was too heavy, a disastrous derailment was apt to occur. By 1876, however, three-quarters of American trains used air brakes, and in 1893 the American Railroad Safety Appliance Act made air brakes and automatic couplers mandatory.[36]

Reporting an accident on the Erie Railroad in 1874, *Frank Leslie's Illustrated Newspaper* included a dramatic wood engraving, "Danger Signal on the Erie Railway . . . Down Brakes!" (fig. 8), in which trainmen leaning in opposite directions anxiously try to stop the speeding train. The brakeman in the smoke-filled locomotive pulls with both arms at the brake while a second trainman tilts his body precariously out the cab doorway into the smoky darkness, pitting his human energies against a seemingly autonomous technology threatening to rush out of control.

The nineteenth-century experience of breaking frame also meant adjusting to new technologies that had a dramatic impact on the way people defined their sense of human identity. New machines with their often gigantic size dramatically shifted perceptions of human size, stature, and scale. Artists' oversize images of mammoth machines suggested the century's inflated pride in technological achievement but also indicated fears of being dwarfed and overpowered by the same technologies.

8 "Danger Signal on the Erie Railway—The Torpedo—'Down Brakes!'" *Frank Leslie's Illustrated Newspaper*, June 20, 1874.

Nineteenth-century art revealed the tragicomic tensions of people attempting to adapt to the imperatives of standardized machines, machines such as the railroad whose speed and schedules sometimes threatened human comfort and safety. Wolfgang Schivelbush in *The Railway Journey* persuasively documented nineteenth-century physicians' concerns about the painful physical and psychological effects of railroad accidents and the impact of riding in jolting machines, but it was nineteenth-century artists who provided some of the most vivid illustrations of how travel anxieties and traumas affected the psyche and the human frame.[37]

Technology also had other unsettling effects on the human psyche and frame, seen in artists' images of mechanized human figures. Audiences in eighteenth-century France had been delighted by mechanical human automatons covered with clothing and hair (see fig. 31), but to critics in industrial societies, including Thomas Carlyle, Karl Marx, and John Ruskin, there was nothing amusing about the prospect of impersonal humans imitating the rote actions of machines. The eighteenth-century automatons were convincing simulacra of the human form—what French sociologist Jean Baudrillard has called a counterfeit, a "perfect double" of the human being.[38] But in nineteenth-century industrialized societies, the eighteenth-century conception of a machine imitating a human, as Baudrillard argues, often became chillingly reversed: humans were transformed into robotic analogues of machines.

Robert Seymour's engraving *Locomotion* (see fig. 30), showing a gentleman leisurely strolling with the aid of his steam-powered legs, suggests this lurking anxiety that people were becoming increasingly mechanized in body and mind—in danger of becoming machine-dependent and mimicking machines. While eighteenth-century craftsmen covered the mechanical parts of their automatons to simulate the look of real human beings, nineteenth-century British caricaturists including Seymour highlighted and exposed the mechanical prostheses that powered their machine-like humans. Seymour's figures were segmented and disjointed rather than integrated and cloaked with a seamless facade, and this quality of disjointedness became a source of humor in his work—suggesting the theme of bodily and psychic fracturing that haunted the century.[39] (Seymour's exposure of mechanical parts had a parallel in architecture—with a difference. Nineteenth-century architects including Louis Sullivan covered the new technology of steel-frame skeletal construction with a facade of ornamented masonry. But with the legitimizing of a machine aesthetic, twentieth-century modernist architects, celebrating technology rather than satirizing it, relished the frank exposure of steel-frame construction and covered their building frameworks only with a transparent glass "skin.")

By the end of the nineteenth century, however, artists were creating somewhat ambivalent views of the man-machine in which workers appeared fused with circular emblems of technology including factory wheels, cylinders, and cogs (see fig. 38). These images in art, as will be seen, suggested both reconciliation and separation—both an intimate connection with technology and a loss of self.

Through their images of exploding steam boilers, towering machines, and mechanized humans, nineteenth-century artists thus skillfully articulated and clarified the century's contradictory views toward technology, capturing the sense of proud achievement as well as undercurrents of skepticism and fear. While these artists were adept at capturing the experience of breaking frame, they also helped contain the radiating, centrifugal forces of explosive technology within the coherent, order-making boundaries of an aesthetic frame.

THE themes of fracturing and integration seen in nineteenth-century artists' views of technology also extended to the world of manufactured objects. As industries flooded the market with factory-made versions of the decorative arts, this too became a source of disruption and anxiety. The eighteenth century had already engaged in mass-producing stamped metals and ornamental silverplate, but the 1840s brought a series of new technological means for creating fine-art reproductions, decorative art reproductions, and imitations. Capitalizing on new materials, technologies, and mechanized modes of production, nineteenth-century manufacturers jolted the art world by introducing a heavy influx of the "imitative arts," mass-produced, machine-stamped, plated, and cast imitations and reproductions of handcrafted original household furnishings, sculpture, and other decorative objects. Adding to the controversy, they touted their new factory-made imitations as superior to more traditional, hand-created works of decorative art.

While achieving commercial success, the new technologies set off century-long debates on both sides of the Atlantic about the aesthetic quality of contemporary industrial design and the desirability of making imitative copies widely available to a middle-class market. To British critics including John Ruskin, the factory-made architectural ornamentation, fancifully decorated cast-iron stoves, and ornament-heavy plated dinnerware were abominations, affronts to authentic art. Worse, they were an assault on cherished aesthetic and social traditions. But to other observers and manufacturers, the century's challenge lay in bridging the gap between the new technologies and the arts traditions they had broken, in finding a way—largely through a choice of ornamental strategies, through improved industrial design, and through professional design-school education—to integrate these new decorative arrivals into the social and aesthetic milieu.

These themes of fragmentation and integration that wove their way through the world of nineteenth-century industrial design are a second major focus in this book. Chapters 3–6 explore the often contentious debates that swirled around the dramatic introduction of the imitative arts and the new, ornamented, factory-made machines. These chapters look, too, at efforts by nineteenth-century art critics, designers, and manufacturers to reconcile art and industry, and to discover a suitable design aesthetic for the new products of industry. Specifically, these chapters view decorative cast iron, electroplating, and steam engines with architectural frames as, each in its own way, a telling paradigm of the often divisive issues raised by the new factory-made versions of the decorative arts.

The 1840s introduced several technologies that were to challenge and transform aesthetic traditions. Chromolithography, the technique for reproducing oil paintings and watercolors in at least three-color printing, was introduced in America in 1840, and by the 1860s color prints were being mass-produced by steam-powered factory machines.[40] Following Louis Daguerre's introduction of his photographic process in 1839, Britain's William Talbot in 1840 introduced the calotype, a photograph on paper which could be readily reproduced. By the 1850s, the wet-plate process made multiple photographic copies even easier to create, profoundly challenging notions of realism, originality, and the look of art itself.[41]

During the 1840s, British and American manufacturers also introduced new versions of the imitative, or decorative, arts: the electroplated, electrotyped, and cast-iron wares seen as appalling by some critics but welcomed by a middle-class market eager for emblems of elegance and status. In 1840, the Coalbrookdale Company in England produced a sizable catalog advertising its ornamental and domestic cast-iron wares, and the same year the British firm of George Richards Elkington challenged silver and silverplate manufacture by taking out a patent for electroplating. And in 1848, New York manufacturer James Bogardus rocked the American architectural establishment by producing the city's first cast-iron building facade, which imitated the look of hand-carved stone with classical Doric ornamentation.

Witnessing this outpouring of mass-produced imitations, England's Augustus Welby Pugin, John Ruskin, and later William Morris insisted that the worlds of art and industry were fundamentally divided, with little chance for reconciliation or union. In a world blighted by industry, Ruskin concluded, there was no way to produce satisfactory industrial design: "to men surrounded by the depressing and monotonous circumstances of English manufacturing life, depend on it, design is simply impossible."[42]

A number of England's most influential art critics were haunted by this sense of division between manufactured and more traditional arts, for they also saw a wide gap between factory-made imitations and the authentic, handcrafted originals.[43] Joining Ruskin, they also complained about the inherent falsity and deceitfulness of factory-made imitations. Mocking the absurdity of trying to imitate motifs from nature using industrial materials, Henry Cole's reform-minded *Journal of Design and Manufactures*, published between 1849 and 1852, lampooned a cast-iron chair made to look like bamboo: "Let the happy nearsighted man, who may be deluded into the idea that these are real bamboo, 'trifles light as air,' attempt to lift one of these perverted three-legged stools, and he will find the difference to his cost."[44]

Also engaging in the century's debates about imitations of handcrafted products, William Morris in "The Aims of Art" complained of another element of divisiveness: the factory division of labor, which severed the organic connection between artisans and their own creations. "The present system," Morris argued, "will not allow him—cannot allow him—to produce works of art." In contrast to the handcrafted work of fourteenth-century artisans, current factory-made imitations were merely "makeshift" art—"this gibbering ghost of the real thing."[45]

While British critics complained about the imitative arts on aesthetic grounds, their language also suggested that they feared that the new imitations were blurring social class distinctions as more people could afford the look (if not the substance) of aristocracy. Indeed, some social observers were pained if not amused by the pretentious longing for status by an upwardly mobile middle class—the type of pretension satirized by Charles Dickens in his novel *Our Mutual Friend* (1865): the two aptly named characters Mr. and Mrs. Veneering were "bran-new people in a bran-new house in a bran-new quarter of London. . . . All their furniture was new, all their friends were new, all their servants were new, their plate was new." The Veneering family and their silverware alike were too recently arrived: "And what was observable in the furniture, was observable in the Veneerings—the surface smelt a little too much of the workshop and was a trifle sticky."[46]

The language of nineteenth-century aesthetic discourse was filled with references to elevation and degradation. Some critics argued that widely available, tasteful examples of the fine and decorative arts would improve public taste and thus be uplifting, but British critics, concerned with protecting aesthetic and social traditions, bemoaned what they saw as the degrading effect of the imitative arts. In the article "Shams and Imitations" published in Cole's *Journal of Design and Manufactures* in 1851, there is language of social class analysis in an aesthetic appraisal of the new imitative arts. Denouncing pretentious or "sham" imitations, the journal proclaimed that "in art, as in morals and politics, a sham is always despicable in the long run, whatever may be its temporary success. . . . A sham being always the pretence of something else, must necessarily be inferior to the thing itself: it can never get higher than meanness, and in many cases—in manufactures especially—it has a tendency to degrade and lessen the appreciation of the superior thing imitated." Referring to machine-made woven fabrics, the journal insisted: "The value of the original is deteriorated by the debased and vulgar affectations of it."[47] The journal's obvious distaste for overreaching in design—its language of superiority and inferiority, debasement, and "vulgar affectations"—all suggested that the new imitations were threatening not only aesthetic conventions but social hierarchies as well.

While British critics, enamored of England's medieval craft traditions and sensitive to social class distinctions, often ostracized the factory-made imitative arts, American commentators—living in a young country with upwardly mobile aspirations—often applauded manufacturers' efforts to integrate the new technologies into their new society and make aristocratic-looking tableware and decorative objects more widely available at a lower cost.[48] The same forces were at work in the introduction of cast-iron architectural ornament. Unfettered by entrenched architectural traditions, looking for quick ways to assemble new building fronts, and eager for architectural respectability, Americans in cities like Chicago, St. Louis, and New York were willing to try out new ornamental cast-iron building facades that imitated the look of hand-carved stone.

Both American and British critics, though, complained about the poor state of contemporary manufacturing design. Increasingly, the focus turned to finding a new design aesthetic appropriate for industrial products. There were efforts, too, to offer artists who had been challenged if not displaced by the imitative arts an increasing role in the shaping of industrial design. One of the problems, argued William Morris in his lecture "Technical Instruction" (1882), was the disjunction between the artists hired to create aesthetic designs for manufacturers and the "technical designers" engaged in translating these designs for industrial production.[49] Too often the transfer of designs produced woeful results.

While Ruskin and Morris, among others, saw themselves living in a time when cherished traditions were breaking apart, splintering, polarizing—a world of design in which art and factory manufactures were often hopelessly at odds—there were important movements afoot during the nineteenth century to ease the sense of explosive disruption and division. Manufacturers, art critics, designers, and art educators developed several strategies to close the gap between art and technology, manufacturing and design.

Manufacturers working to unite art and industry produced factory machinery decorated with painted and architectural ornament. Steam engines and many other machines were themselves new arrivals in the cultural milieu, and designers faced with the challenge of finding an appropriate style often ornamented them with decorative striping, stenciled patterns, and architecturalized cast-iron frames with classical columns or other ornamental details (see fig. 58)—a practice that would begin to be critically scrutinized during the last decades of the century.

During the twentieth century, these efforts at industrial ornamentation have been commonly viewed as reflecting the nineteenth century's love of ornament itself.[50] John Kouwenhoven's influential study *Made in America* (1948) linked America's practice of decorating machines to the country's wish to continue the European "cultivated tradition" and to the fact that early machines were often made of wood by cabinetmakers accustomed to ornamental embellishment.[51] Historian John Kasson in *Civilizing the Machine* (1976) has suggested that ornament added sales appeal to mechanically mysterious new technologies and was associated with quintessential American republican values: Americans wished to see utilitarian objects and machinery as inherently beautiful and even artistic. Designers decorated machines to help "assimilate the machine into republican civilization, to signify its honored status in the life of the nation, and to enhance its reputation as art."[52]

But, as will be explored later, the practice of ornamenting machines had other important cultural meanings as well. By architecturalizing factory and domestic steam engine frames, camouflaging them with the ornamental motifs popular during the century, manufacturers could help ease public anxieties about disruptive new technologies. Engines designed as classical temples of antiquity lent the stasis and calm of a harmonious frame to an unsettled world of technological change. By ornamenting machines with neoclassical, Egyptian, or Gothic revival styles, some of the most popular design idioms of the day, manufacturers also found a way to legitimize the new arrivals by associating them with aristocratic tradition. Like the nouveau riche, middle-class members of Gilded Age and Victorian society, new technologies were brash upstarts who had moved into the neighborhood and were suspected of vulgarity and lack of breeding. Cloaked in their historic dress, the machines acquired cachet, appearing genteel and stately, demanding dignity and respect.

Though nineteenth-century British and American architecturalized steam engines were often admired for their towering and magisterial presence, and although British and American consumers were eagerly purchasing the new cast-iron furniture and electroplated tableware, the factory-made decorative arts did not fare particularly well in the eyes of art critics. Surveying the design of decorative cast iron in England, Charles Eastlake in his influential *Hints on Household Taste* (1868) wondered why "the British public go on buying such trashy articles." Eastlake's book, which was widely reprinted in both England and America, ridiculed the ornament on electroplated wares, where he found "a rose, a tulip, and an apple respectively doing duty as the handle of a tea-pot lid" and a butter cooler "surmounted by that inevitable cow." If only artisans were given a "sound education," he argued, "we might hope for something better than the everlasting palm trees, camels, and equestrian groups" that gratified contemporary taste. "In such objects," he complained, "the designer does little more than model more or less correctly after nature." "This," he concluded, "is imitation but not *design* in the artistic sense of the word."[53]

To counter what they considered to be the abysmally low aesthetic quality of industrial products, design reformers tried to forge a closer link between art and manufacture and also improve design education. Intent on improving the quality of British design, Henry Cole, using the pseudonym Felix Summerly, started a firm producing his own "art manufactures" in 1847, employing the talents of noted artists and designers. Britain's *Art Journal* and the American edition of the *Art Journal* also encouraged manufacturers to increase their hiring of professional artists to design their wares. Writing in the *Journal of Design and Manufactures* in 1851, Matthew Digby Wyatt urged artists to become more knowledgeable about industrial materials and manufacturing methods so their designs would be in harmony with the special requirements of manufacture.[54]

During the 1870s, the practice of ornamenting and colorfully painting machines and tools (particularly in America) was being increasingly scrutinized by art critics, industrial designers, and engineers alike who often attributed the poor quality of industrial design, and the misuse of ornamentation, to faulty systems of art education. Too often, they argued, designers had inappropriately adapted an aesthetic for handicraft to the mass-produced imitative arts. As the reformers struggled to find the right aesthetic grammar of ornament, they noted that all too often there was a confusion of languages: ornament considered suitable for handcrafted silver or marble, for example, might not be suitable for the new machine-stamped and plated metals. Improving the education of designers, the critics argued, would help them forge a new aesthetic appropriate to the functional requirements of manufacturing and appropriate to the new industrial ethos itself. The nature of this aesthetic, though, continued to be hotly debated.[55]

Parliamentary debates about the state of British industrial design had already resulted in the founding of a national School of Design in London in 1837 and regional design schools throughout Britain. In the United States, educators including Walter Smith, superintendent for art instruction for the state of Massachusetts, made similar arguments for improved art education beginning in the 1870s, and design schools were established throughout the country by the end of the century.

Another way to create closer ties between art and manufacturing, and to help heal the perceived separation between fine artists and industrial designers, was to urge the professionalization of industrial designers—a goal championed by Christopher Dresser, one of England's most respected designers. Reflecting the nineteenth-century design themes of division and integration, degradation and ennoblement, the low and the high, Dresser's lectures identified a profound split between pictorial artists and artists engaged in "applied art," the latter often relegated to low status. Dresser's preoccupation was with heightening the status of designers and industrial "ornamentalists." As he wrote in "Hindrances to the Progress of Applied Art" (1872), design schools often had no one who honored the art of ornamentation and infused into the students "a feeling of its nobleness and greatness."[56]

Engineers had made very occasional references to design reform as early as the 1820s, but beginning in the 1870s, engineering journals in both England and America began responding to contemporary debate about art education and industrial design, calling not only for improvements in industrial education but also for a change in the prevailing industrial design aesthetic. During the 1880s, a few technical journals were becoming outspoken about the idea of abandoning ornament in favor of an aesthetic of functional simplicity.

In an article published in 1881, *Scientific American* took a dim view of artists engaged in designing "'artistic' machines and tools," and praised those who designed machines "best adapted to perform the work required of them with the least outlay of material," without "disregarding shapeliness and harmony of proportion."[57]

The emphasis on subordinating ornamentation to utility had been a recurrent viewpoint among nineteenth-century design reformers, heard often in the pages of Cole's *Journal of Design and Manufactures*, where the complaint had been that too often utility was sacrificed to ornament. Christopher Dresser in 1872 had also argued for the unification of the two, writing that "the most perfect beauty may be combined with the most complete utility."[58]

The call for design simplicity in late nineteenth-century science and engineering journals may well have been a recognition, or affirmation, of the plain, unornamented vernacular American styles that had been present, as John Kouwenhoven and others have reminded us, in various guises in American machine design all along. The praise given to Dresser's elegant electroplated designs, and manufacturers' willingness to entertain the idea of discarding ornament in favor of functional design during the late nineteenth century, also illustrated an irony of design history: the fact that plain-style, functional designs were being touted as palatable if not preferable was due, in some measure, to manufacturers' successful use of ornament to help dignify and legitimize the machine.

After nearly a century of struggle to dignify and integrate the imitative arts and disruptive new technologies into the prevailing social framework, a struggle that often entailed the use of ornamental machine frames, engineers and designers could argue for the pristine beauty of plain, unadorned machine forms, suggesting that these machines no longer needed elaborate ornament to heighten their status and assure their legitimacy. American engineer Henri Haber, writing in *Engineering News* in 1877, argued that too much ornamental painting "may destroy the impressions of repose, earnestness, and dignity belonging purely to works of engineering."[59]

By the end of the century, the neoclassical columns, entablatures, and other historical detailing that had given status to nineteenth-century industrial designs were largely being replaced by a style of classical simplicity which became the dominant aesthetic of twentieth-century modernism. (But as another of the continuing ironies of design history, the success of the classically simple, machine-inspired modernist aesthetic in the twentieth century would lead postmodern architects to complain about the pervasive presence of unornamented designs. In the 1970s and after, American architects including Michael Graves and Charles Moore reintroduced historicized ornamentation, particularly stylistic quotations of classical ornament, as a way to humanize the cold, mechanistic look of modernist design.)

Bringing together some of the central themes that characterized design debates in the previous century, American architect Frank Lloyd Wright in his lecture "The Art and Craft of the Machine," presented in Chicago in 1901, engagingly articulated the nascent machine aesthetic as he discussed the changes that were reshaping the world of architecture and industrial design. His lecture was a remarkable testament signaling a time of transition: he scathingly denounced the aesthetic trauma of the imitative arts, which had haunted the nineteenth century, and also looked forward to the emerging machine aesthetic, which would transform twentieth-century design.

Looking inside contemporary homes with their imitative, pseudo-historical styles—the "Chateaux, Manor Houses, Venetian Palaces, Feudal Castles, and Queen Anne Cottages"—he saw only the blighting influence of the imitative arts, the "machine-made copies of handicraft originals." "In fact," he wrote, "unless you, the householder, are fortunate indeed, possessed of extraordinary taste and opportunity, all you possess is in some degree a machine-made example of vitiated handicraft, imitation antique furniture made antique by the Machine, itself of all abominations the most abominable."[60]

But for all his complaints, Wright also idealistically envisioned the transforming, even poetic effects of modern technologies and the imitative arts: "The modern processes of casting in metal" were "approaching perfection, capable of perpetuating the imagery of the most vividly poetic mind without hindrance—putting permanence and grace within reach of everyone." He saw the advantages of machines for woodwork, machines that had "emancipated beauties of wood nature, making possible, without waste, beautiful surface treatments and clean strong forms."[61]

These "clean strong forms" exemplified the promise of the machine, a promise that was often wasted during his age, when debased designs produced "a riot of aesthetic murder." Wright's own early architecture looked to other stylistic influences, including Japanese and Shingle Style architecture, but he clearly appreciated the spare abstract forms of the machine. "The gift of the Machine," he concluded near the end of his lecture, lay in its "great possibilities of simplicity."[62]

As Wright emphasized, nineteenth-century artists and industrial designers had mirrored, exacerbated, and in some ways had also begun to transform the technological tensions of the age. They recognized a condition of splintering and fragmentation, even as they sought some form of rapprochement with technology. Their practice of ornamenting the products of industry and machines was, in a sense, a way of accommodating and easing the disruptive impact of new technologies. Ultimately, the new aesthetic of design simplicity, which they advocated toward the end of the century, reflected the success of their own efforts to dignify machine technology and to heighten the status and professional training of the manufacturing designer.

Inspired by the clean lines of machine forms, they turned to a design aesthetic of classical clarity and abstract, geometric simplicity—an aesthetic that would become the central idiom for modernity in the new machine age. Reasserting the century's twin themes of fragmentation and fusion, splintering and integration, their starkly simple, unornamented versions of the classicized machine offered a way of creating a vital link between technology and art—and a coherent framework for a world still in the midst of fragmenting, disruptive change, a world that would become even more intensely engaged with explosively exciting, and explosively dangerous, new technologies.

Chapter 1

The Traumas of Transport in Nineteenth-Century Art

In its July 8, 1808, issue, the London *Times* announced that inventor Richard Trevithick was preparing to race his new experimental steam engine or locomotive in a contest "against any mare, horse, or gelding" at an October 24 meet at Newmarket, with the engine viewed as the favorite. "The extraordinary effects of mechanical powers [are] already known to the world," the *Times* rhapsodized, "but the novelty, singularity, and powerful application against time and speed has created admiration in the minds of every scientific man."[1]

The *Times*'s excitement over Trevithick's high-pressure engine, which was provocatively named *Catch me who can*, captured the early nineteenth-century delight in spirited new technologies—yet as the engine's name tauntingly suggested, these same machines also threatened to speed away, maniacally evading human control. To demonstrate his engine, Trevithick in 1808 set up a circular track and charged five shillings each to passengers who rode in a carriage behind the puffing machine. But in spite of high expectations, *Catch me who can* was unceremoniously derailed on a broken track. Short of patronage and funds, Trevithick discontinued the demonstrations, though he would go on to develop other, more successful models.

The contradictory feelings associated with Trevithick's engine—heightened hopes and the shock of unexpected accidents—typified the dialectic of the age. The heady optimism generated by Robert Fulton's scheduled steamboat service on the Mississippi and the inauguration in 1830 of England's Liverpool and Manchester Railroad, the earliest regularly scheduled passenger railroad, was tempered by pessimism over the sobering news of boiler explosions on steamships and trains which periodically shattered the nineteenth century's pervasive rhetoric of progress.

Passenger responses to the new modes of transport were often contradictory. While some railroad passengers were exhilarated by traveling at new speeds, others were terrified, reporting feelings of disorientation and foreboding. British actress Fanny Kemble, invited to ride as a guest passenger with George Stephenson on a preliminary run of his locomotive the *Rocket* on August 26, 1830, described her experience traveling at thirty-five miles an hour: "You cannot conceive what that sensation of cutting the air was.... When I closed my eyes this sensation of flying was quite delightful, and strange beyond description." To Kemble, the train was a "tame dragon" that "flew panting along his iron pathway."[2] But British writer Thomas Carlyle somberly saw the railroad as an untamed, diabolical technology. Describing a train trip in a letter to John Carlyle dated 1839, he wrote: "The whirl through the confused darkness, on those steam wings, was one of the strangest things I have ever experienced—hissing and dashing on, one knew not whither.... Out of one vehicle into another, snorting and roaring we flew: the likest thing to a Faust's flight on the Devil's mantle."[3]

In their paintings, documentary illustrations, and satirical caricatures, nineteenth-century European and American artists produced vivid images of these often conflicting responses to technology. Their work illuminated both the triumphs and the exasperations of breaking frame—of people struggling to reorient themselves to an age of machines. In a few deft and incisive lines, Honoré Daumier could capture the look of comic embarrassment and bewilderment as people attempted to negotiate the unfamiliar territory of a new technology, as in his lithograph of 1855 where hesitant travelers stare anxiously at train carriages, trying to decide which is the safest and terrified at the prospect of entering into the technological unknown (fig. 9).

Painters interested in treating these problematic encounters with technology were rare: perhaps many agreed with poet William Blake, who, having read the first issue of Britain's *Mechanics' Magazine* in 1827, was reported to have exclaimed, "Ah, sir, these things we artists HATE!"[4] But for a small number of artists—primarily caricaturists and illustrators for European and American weeklies—the subject of

32   Breaking Frame

9 Honoré Daumier, "Le Choix du Wagon," *Les Chemins de Fer* (2d ser.). Lithograph. *Le Charivari*, December 1855. Collection of the author.

people negotiating new, sometimes threatening machines held a particular fascination. Among these social commentators, the revolution in transportation brought by experimental steam carriages and passenger-carrying railroads became one of the most popular subjects to illustrate.

In a century devoted to idealized travel prints often commissioned by railroad companies to promote ridership, there was a lesser-known category of prints which focused not on safety but on dangers. Illustrators and satirists, capitalizing in part on the public's appetite for sensationalism, brought to the surface an undercurrent of tension and anxiety associated with travel by steam, presenting macabre images such as Jules Tavernier's illustration of a Rhode Island train crash published in *Harper's Weekly* in 1873, where a rushing train heads toward the grim, ghostly figure of a grinning skeleton, making manifest deeply embedded fears of steam explosions and of runaway vehicles going too fast to be stopped.

The pictorial images of explosions and runaway technologies were not simply artists' fantasies. Nineteenth-century newspapers in Europe and America repeatedly carried stories of exploding boilers on high-pressure engines and steamships. As American engineer James Renwick wrote in his *Treatise on the Steam Engine* in 1848, "The subject of the explosion of steam boilers has recently attracted a great share of public attention."[5] Published investigative reports detailed the grim statistics of a steam boiler explosion on New York's Hague Street in 1850 in which sixty-seven people were killed or injured. In England in 1844, an explosion at Brookes's Flax Mill at Lancashire shot a twenty-seven-foot-long, ten-ton steam boiler into the air; it landed twenty yards away, accompanied by a sound "like that from the discharge of an immense piece of ordnance."[6]

American and European investigators sought to stem the numbers of these explosions. In its report of 1836, Philadelphia's Committee of the Franklin Institute on the Explosion of Steam Boilers described its experiments conducted for the U.S. Treasury Department in which researchers tested boiler cylinders with a gradual increase of pressure, and in 1838 the U.S. Congress passed a series of boiler regulatory laws.[7]

After Robert Fulton began his steamboat runs in New York in 1815 and steamboats were introduced in England, boat explosions became a growing public concern. A fatal explosion on a steamboat near Norwich in England in 1817 prompted British safety legislation mandating supervised boiler construction and the use of two safety valves, but steamship boilers continued to explode. Although England experienced only 77 deaths in twenty-three explosions from 1817 to 1839, Americans suffered a very high number of steam casualties due to a variety of factors including the much greater number of vessels afloat compared with the rest of Europe, the higher tonnage being carried, poorly designed safety valves, and a shortage of competent engineers. The U.S. commissioner of patents reported that in the single year of 1838, 496 people died in fourteen explosions.[8]

Steamboat explosions remained a public concern even as boat owners sought to provide reassurances of safety. The Connecticut River Steamboat Company in its report of 1833 detailed the causes of the explosion on the steamboat *New England* that year but claimed that "this mode of travelling is even now safer on the average than any which is afforded by the ordinary method of conveyance." Nevertheless, a U.S. Senate report in 1848 noted that the question of finding "the best method of preventing the bursting of steam boilers has for a long time agitated the public mind" and complained that "the evil has not been *much diminished*."[9]

Burst boilers on railroads were also a source of public concern, as when the *Charleston Mercury* reported that the *Best Friend of Charleston*, a locomotive built for the South Carolina Railroad which pulled America's first passenger-carrying train on December 25, 1830, was destroyed by a boiler explosion on June 17, 1831, killing the fireman. The South Carolina company suffered more explosions in the years that followed as the boiler of its new engine, the *Buena Vista*, exploded in 1848, and three years later a similar accident destroyed the locomotive *James L. Pettigru*, in both cases killing the engineer and two firemen. Explosions on British railway engines were first reported in England in 1815, and during the period from 1861 to 1870 alone, thirty-nine boiler explosions were reported on British railroads.[10]

Reports of runaway steam vehicles were also published in British newspapers during the 1820s, although it was not uncommon for the newspapers to attempt to soften the impact of these mechanical upheavals. The London *Observer* in 1827 described an experimental steam carriage journey in which the carriage made a runaway descent down a hill until it was propelled violently into a gutter. Having presented an archetypal image of technology out of control, the newspaper concluded optimistically that the event would undoubtedly lead to "suggestions as must, in a very short distance of time, bring steam carriages into constant and beneficial use for the conveyance of both goods and passengers."[11] Yet the satirists' prints of steam explosions which began appearing in England as early as 1829 could be more broadly seen as a metaphor for skepticism and distrust of the new speeds of travel—and a reflection, too, of more subliminal anxieties that social frameworks were speedily being shattered.

Railroad accidents and traumas were rarely, if ever, shown in nineteenth-century paintings, which were more apt to document the dignity and progress of railroads. In J. M. W. Turner's majestically titled *Rain, Steam and Speed—The Great Western Railway* (1844), the innovative, broad-gauge railway locomotive emerging through a swirl of light and color, its red cyclopean lamp aglow, becomes an iconic statement of the nineteenth century's sense of technological imperative. Crossing a bridge during a storm, the locomotive emerges eerily through a miasma of fog and mist. While most of the scene is painted with Turner's characteristic soft, diffuse brush strokes of swirling color, the locomotive's smokestack alone is endowed with clarity and definition, lending legitimacy and authority to the speeding new technology.

Other paintings and prints gave ample evidence of artists' efforts to soften and tame the new technology—to ease its disruptive impact on the natural environment and on human lives. The history of railway art reveals artists' efforts to camouflage technology by embedding machines in the natural landscape. Nineteenth-century British artists were hired by railroad companies to assuage public fears by making safe, sanitized images of trains traversing a tranquil countryside, free from the lurking danger of explosions and accidents.[12] In the idealized view of the ubiquitous and immensely popular nineteenth-century British railroad prints, train travelers often appear as distant, minute figures whose responses to the railroad are not shown.

American artists presented similar images of the new technology. The few American painters who recorded the presence of "the machine in the garden," to use Leo Marx's central metaphor for the incursion of technologies such as the railroad into the nineteenth-century mythic pastoralism, often presented the machine as a benign presence, absorbed without alarm into the texture of the terrain and posing no threat to humans. As Marx argued, "Americans had little difficulty in reconciling their passion for machine power with the immensely popular Jeffersonian ideal of rural peace, simplicity, and contentment."[13] Perhaps the best-known American railway painting, *The Lackawanna Valley* (1855) by George Inness (fig. 10), commissioned by the Lackawanna Railroad Corporation, incorporated the potentially disruptive machine into the lush greenery of rural Pennsylvania. If Thoreau at Walden Pond was momentarily startled by the sound of the railroad, Inness's view presents a relaxed observer in the field; the painting ultimately remains a commercial vision, taking a distant, idealized perspective of railroad travel in which the machine has little disruptive impact.[14]

More indicative of the problematic relation of the machine to the American natural and human landscape is the painting *Steamboat on the Ohio* by Thomas Anshutz (1896), in which a boater in a skiff and nude figures of boy swimmers stand by the river shore as silent witnesses to the passing of a steamboat, which has assumed the right-of-way. This is not the destructive steamboat that suddenly disrupts the tranquillity of Huckleberry Finn and Jim on their raft in Twain's novel and sends them crashing into the Mississippi. Anshutz's steamboat has its own legitimacy and imperative, its smokestacks merging with the smoke of factories on the river's far shore. Its passage is assertive but temporary, disrupting the natural scene which will, in

10 George Inness, *The Lackawanna Valley* (1855). Oil on canvas. National Gallery of Art, Washington; Gift of Mrs. Huttleston Rogers.

time, absorb the intrusion and return the river back to the skiff and swimmers.

GRAPHIC SATIRES AND STEAM TRANSPORT

It was European caricaturists, much more than American artists, who rejected a sanguine view of steam transportation. The pictorial convention of using radiating lines to depict steam explosions in the prints of British artists Robert Seymour, Hugh Hughes, and William Heath charted the tremors of technological change. British caricaturists—ever alert to topical social issues and the public's preoccupations and anxieties—spoofed the new modes of steam travel by showing people being blown apart or enjoying an array of fantasy steam-travel inventions.

In the eighteenth century there was a large volume of popular satiric prints by artists such as Hogarth, but copper-engraved satiric prints by such artists as James Gillray and Thomas Rowlandson were often relatively expensive and considered collectors' items. By the 1820s, however, British print sellers including the leading print seller, Thomas McLean of Haymarket in London, were selling larger, cheaper editions as prices fell with the spread of lithography and wood engraving. The development of lithography in 1798 by the Bavarian Aloys Senefelder had opened up the possibility of quickly executed prints, and this new medium, along with wood and copper engraving, provided the tools for producing comically exaggerated images of human entanglements with technology which had an immediate, direct appeal.[15] Pictorial satire became an important strategy for subduing technology through laughter: artists' engravings and lithographs appeared to be less a manifestation of gratuitous venom than of particularly keen sensitivity to the absurd, awkward, and dangerous aspects of human encounters with mechanization.

The prospect of steam transportation had early caught the imagination of Erasmus Darwin, grandfather of Charles and friend of James Watt and Matthew Boulton, who in his poem *The Botanic Garden* (1791) prophesied the steam age in reverential language:

> Soon shall thy arm, UNCONQUER'D steam! afar
> Drag the slow barge, or drive the rapid car;
> Or on wide-waving wings expanded bear
> The flying-chariot through the fields of air.[16]

Darwin's optimistic expectations were borne out, at least in quixotic nineteenth-century experiments with steam-driven flying machines, but British satirists shared little of his enthusiasm. William Heath (1795–1840), who for a period signed his prints "Paul Pry" accompanied by a small figure of the stage character with the same name, ridiculed the public's preoccupation with new modes of transport.

Heath began as a watercolorist, painting portraits and military subjects including aquatints of the Battle of Waterloo, but after 1820 he turned to caricature as the demand for military prints lessened. Becoming London's leading caricaturist from 1828 to 1830, he was a major illustrator for *The Looking Glass*, a caricature monthly. By August 1830, though, Heath, whose prints had been highly popular, was out of favor, a victim in part of the increasing preference for lithographs over etchings. He was ousted from *The Looking Glass*, and his etchings were replaced by the lithographs of Robert Seymour, leading Heath to turn to book illustration and topographical prints.[17]

Heath's sardonic images were particularly skeptical of contemporary aerial experiments and steam carriages, which appear in his engravings as fanciful and farcical devices destined to reduce rather than heighten human stature, making people look foolish and ridiculous. Hot-air balloon travel had, in fact, become a popular novelty following the first flights of the Montgolfier brothers in France in 1786, and Heath's two versions of *The March of Intellect* (1829) lampoon balloon mania with images of a "Balloon Brigade" of "Pegasern Lancers"—cavalry soldiers mounted on winged horses with balloons on each hoof. Other fantasy transports include a finned and flying "Aerial Ship" built "on the model of the Flying Dutchman" as well as a ship in water propelled by a kite flying overhead.

The ideas of a kite-propelled ship and carriage were actually discussed in a London *Observer* article of 1827 on "New Modes of Travel" headlined "Aeropleustics, or Navigation in the Air." The *Observer* reported that tales of kite-driven carriages traveling at twenty miles per hour were true, and that patents had been taken out in London and Paris by inventor George Pocock, who reported his kite-propelled sailing trial. Pocock's account of the sailing experiment, which appeared in the *Observer* article, included a telling account of runaway technology: "While scudding along under our buoyant sails, an oar had dropped overboard, and having no means of stopping the boat, we traversed to the nearest shore."[18] In William Heath's comic engraving of a kite-driven craft, a wary ship's captain braces against the pull of a kite and struggles to moor the boat, anxiously calling to his boathand, "Tell the Captain of the Kite to take in a reef."

"The March of Intellect" was, for a time, a popular phrase in British caricature and literature after 1825, often suggesting a sardonic view of new steam inventions and movements for working-class education. Heath's second version of *The March of Intellect* (1829, fig. 11), again spoofed transportation experiments of his day, as in his image of a woman astride a three-wheel chair being propelled through the air by a kite, bellows, and a propeller, along with another fanciful proposal of carriages propelled by steam through vacuum tubes.

In his print, Heath also took particular delight in satirizing one of the more serious inventions of the era: the steam carriage, such as the type developed by British inventor Goldsworthy Gurney, who made his initial experimental runs starting in 1827 and carried passengers for three months on a Gloucester-Cheltenham service in 1831 (fig. 12). Wayward steam carriages were particulary suitable targets for satire and highlighted the nineteenth century's love-hate relationship toward technology. The pictorial image of a steam carriage

11 William Heath, *The March of Intellect* (1829). Engraving (detail). Ironbridge Gorge Museum Trust, Elton Collection.

became a central emblem of the twin anxieties haunting technology in art: fears of explosions and unstoppable machines speeding out of control.

The earliest steam carriages were designed by French military engineer Nicolas Cugnot, who in 1770–71 developed a three-wheel high-pressure steam machine with a large copper boiler that over-

12 Gurney's steam coach (1827). Ironbridge Gorge Museum Trust, Elton Collection.

hung the front wheel. Further steam carriage experiments were carried out by William Murdock, the assistant to Matthew Boulton and James Watt, but Murdock's model was opposed by Watt, who was skeptical of high-pressure steam. British inventor Richard Trevithick was later credited with developing early successful high-pressure steam locomotives in 1801 that eliminated the need for steam condensers, yet the dangers of high-pressure steam were dramatized in 1803 when one of his stationary engines at Greenwich exploded, killing four people.

In 1827, the London *Observer* enthusiastically reported the first public appearance of Gurney's steam carriage, which made a trial run in Regent's Park on December 6. The carriage consisted of a boiler, U-shaped furnaces that burned coke or charcoal, cylinders, and pistons, all of which generated eight to twelve horsepower. It was designed to carry twenty-one passengers, fifteen sitting outside and six inside. The body was shaped like a stagecoach but was larger (fifteen feet long) and weighed one and a half tons. But as noted by later historians of the steam engine, the carriage's butt-welded tubes, along which gases were directed, had the almost ineradicable defect of bursting.

The *Observer*'s report of the 1827 trial run in Regent's Park encapsulated contemporary enthusiasms, tensions, and rationalizations about technological change. Describing the development of the steam carriage, the paper extolled "the rapidity of its progress" and "its perfect safety," arguing that "in a very short time it will receive the sanction of the public," but also acknowledging that "many prejudices will, of course, have to be combated," for "at the first mention of a *steam boiler* attached to a coach we are not surprised that the timid should take alarm." Indeed, *Observer* issues of the same year contained several accounts of steam explosions. In an effort to assuage readers' doubts about the new steam carriage, the newspaper took a rhetorical strategy that would increasingly be used by writers and artists to tame technology: it transformed the machine by humanizing it, drawing on an anthropomorphic metaphor to make the strange seem familiar. Gurney, said to be trained as a medical man, had with his carriage "actually made the construction of the human body, and of animals in general, the model of his invention." In the *Observer*'s description of machine as body, reservoirs for steam and water became "the heart of the steam apparatus," the lower boiler pipes were arteries, the upper pipes veins, and water was equated with blood.[19]

Personal accounts by those who actually traveled in experimental steam carriages also reflected a mixture of anxiety and optimism. In his memoirs, Lieutenant General Herbert Taylor, private secretary to King George III, included two reports by Sir J. Willoughby Gordon and Lieutenant Colonel Sir Charles Dance, who traveled on separate Gurney steam carriage rides from London to Bath in 1829. Gordon, describing the carriage's steering mechanism (a horizontal wheel in front of the seat, connected to the axle and front wheels), emphasized the carriage's controllability, assuring the reader that the carriage "may be directed with greater precision than can any carriage drawn by horses under the direction of a coachman" and that "the machine can be stopped instantly."[20]

A less sanguine account was given by Sir Charles Dance (who later, in 1831, took over Gurney's carriage service on the nine-mile Gloucester to Cheltenham run, although he was soon forced to stop the service due to opposition from owners of horse-drawn carriages and innkeepers). Dance reported his traumatic trial run to Bath on July 28, 1829, during which the carriage, traveling at ten miles per hour, carried four persons in a barouche attached to a boiler and was accompanied by an advance carriage and one behind containing coke for the steam boiler and carrying some of the engineers, including Gurney. Dance's account began with reassurances of the coach's safety but concluded with a revealing demonstration of broken frames.

The trouble the carriage encountered was not mechanical but human. Arriving at the town of Melksham, twelve miles from Bath, the passengers discovered a town fair in progress and moved the carriage slowly, but the appearance of the machine created a stunning outbreak of public hostility. Dance's own response was one of bewilderment: "Unfortunately from some cause or another the people here had taken a dislike to the steam-carriage, and after abusing us shamefully, attacked us with stones and flints, and after having wounded the stoker and another engineer severely on their heads (the former being knocked out of the carriage into the road) a violent scuffle took place between us."[21] The carriage was later returned to a repair shop and towed by horses to Bath. The violent public outburst revealed a strong undercurrent of suspicion and fear toward new technologies.

What ultimately stopped the steam carriage in Britain, however, was economic and political opposition, as well as mechanical difficulties. The carriages were opposed by business and agricultural interests, who viewed the machines as a threat to systems of horse-drawn carriages and canal barges, and who ensured the carriage's demise through parliamentary laws creating prohibitive road tolls. Contemporary reports also told of mechanical failures and breakdowns caused by the difficulty of carrying hot, heavy, vibrating engines, by heavy water consumption needs on long trips, and by broken axles and exploding boilers.[22]

The carriage attack at Melksham was reminiscent of the Luddites' machine-smashing rampages from 1811 to 1813. Nineteenth-century caricaturists, however, showed their own skepticism of the steam carriage not through violence but through ridicule. British artists including George and Robert Cruikshank, William Heath, and Robert Seymour viewed the steam carriage as a machine that threatened to break social frameworks by displacing the horse-drawn carriage, polluting the atmosphere, and worse, causing explosive accidents that sent travelers' bodies flying.

The steam carriage's economic threat to conventional horse-drawn carriages was bitingly satirized by George Cruikshank (1792–1878), who had established himself as the heir to James Gillray and was the leading caricaturist of the British Regency period. His engraving

Horses 'Going to the Dogs' (1829) burlesqued contemporary concerns about the displacement of horse-drawn vehicles. In the engraving, a group of talking horses and dogs standing on a rural hill discuss the passage of a steam carriage on the road below: "Well dash my wig if that isn't the rummest go I ever saw!!" exclaims one horse as another blind horse exclaims, "A coach without horses!! nonsense—come, come, Master Dobbin you are 'Trotting.'" Meanwhile, a philosophical dog, "Wagtail," happily considers the imminent availability of horse meat.

But the deeper sources of worry posed by the steam carriage were its speed and its potential for dangerous explosion. William Heath's second engraving of *The March of Intellect* (see fig. 11) made repeated jabs at the steam carriage's ability to shorten travel time and spatial distance. Heath's satiric steam wagon advertises "London–Bath in six hours," which only somewhat exaggerates an actual London–Bath steam carriage excursion made by Gurney the same year that traveled the eighty-four miles in ten hours with stops, averaging fifteen miles per hour. The images in Heath's engraving reflect the artist's underlying uneasiness about the machine's speed and the relentless momentum of new technologies: a steam vehicle in the shape of a large horse named "Velocity" bears the legend "No stopage on the road." Heath's hyperbole was also aimed at travel inventions that made the world seem suddenly a disorientingly smaller place. His fantastic aerial devices include a grand vacuum tube going "Direct to Bengal," a batlike ornithopter carrying convicts to Australia, an improbable suspension bridge to Capetown, and prophetically, a flying postman.

The disorientation caused by a contracted planet was paralleled by fears of a world made increasingly dense and crowded through technology. British and Austrian caricaturists prophesied a future of traffic jams and polluted skies. *A View in White Chapel Road, 1831* (1828) by Britain's Henry Alken and John Leech's *Hyde Park As It Will Be* (c. 1848, fig. 13) predict a London packed with steam-powered vehicles (bearing comically sinister names such as "The Dreadful Vengeance" and "The Infernal Defiance") that emit clouds of black smoke while carrying cheerful passengers undaunted by the congestion and smoke. (A contemporaneous Austrian engraving, *Steam Wagons in the Year 1942 in Vienna*, envisions a much more chaotic scene, a time during the next century when colliding steam carriages spew forth fire and smoke as startled passengers are unceremoniously toppled into the street.)

13 John Leech, *Hyde Park As It Will Be* (c. 1848). Lithograph. Collection of the author.

Such images became a central metaphor for the disjunctive impact of technological change. In the early 1820s satirists had spoofed the comic aspects of conventional carriage accidents. Travelers on horse-drawn carriages were subjected not only to a jarring ride but also to the occasional trauma of an overturned vehicle, as seen in the ironically titled *Comforts of a Cabriolet* drawn by M. Edgerton and etched by George Cruikshank (1821), where a carriage mishap caused by a rearing horse sends passengers flying high. Beginning in 1825, however, the introduction of steam carriages, with their potential for violent explosions, offered caricaturists a new target, even as carriage designers were working to develop improved boilers to prevent such accidents. Artists quickly developed pictorial conventions to convey the explosive destructiveness of burst boilers, including exaggerated violence, flying bits of debris, radiating lines, and human bodies thrust high in the air. In *New Principles or the March of Invention* published by London's Thomas McLean, an exploding steam coach violently ejects frock-coated passengers while relaxed travelers flying in a rainbow-striped balloon overhead, unaware of the catastrophe below, comment complacently, "We'll take a flight to Heaven tonight, and leave the dull earth behind us."

It was British satirist and book illustrator Robert Seymour who created some of the century's most vivid prints of steam explosions. Originally a painter in oil, Seymour turned to caricature, producing popular folios of copper engravings including *Humorous Sketches* (1834, 1836), and became best known for his illustrations for the first two issues of Charles Dickens's *Pickwick Papers*. Seymour's life was abruptly truncated when he shot himself in the head in 1836, an act triggered by his disputes over the illustrations for *Pickwick* and with London's *Figaro*, the predecessor of *Punch*. Irritated with Seymour, the *Figaro*'s publisher had printed a series of scurrilous attacks on him which apparently disturbed the sensitive and nervous artist, said to be subject to bouts of depression.[23]

Seymour's caricatures, perhaps due to the artist's own frail temperament, were especially effective at capturing the trauma of fractured technological frames. In his version of a steam carriage explosion, *Going It by Steam* (1829, fig. 14), passengers are hurled into

14 Robert Seymour, *Going It by Steam* (1829). Engraving. Guildhall Library, City of London.

oblivion, while the still-bonneted, severed head of a woman is seen separated from the trunk of her body, which sprawls inside the carriage. Seymour's images of exploding steam carriages with their grotesqueness and violence deftly detail the new technology's deadly force. While other contemporary artists also caricatured exploding steam carriages and fragmented bodies, Seymour best captured the anguished, hair-raising expressions of shocked travelers. The explosive energy of technology in his caricatures rudely invades and shatters the sanctity of seemingly secure enclosures, causing a disruption that is both comic and cruel. In *A Steam Coach Passenger Set Down or an Unexpected Arrival by Steam*, the soaring body of a woman passenger expelled from an exploding steam coach unexpectedly flies into an elegant dining room, upsetting tea table and guests. It is this disruptive element of surprise and uncertainty that endows Seymour's prints with an element of the absurd. For all their comic exaggeration, Seymour's works skillfully capture a nightmare of shifting structures, a world where new technologies are launching an invasion with explosive force.

THE RAILROAD: DANGER, STEAM, AND SPEED

Satires of the steam carriage were brought to an abrupt halt as the carriages were supplanted in the 1830s by much more viable railroad systems. The railroad soon became the dominant image of technology in art in the second half of the nineteenth century, an image that encapsulated the century's technological hopes and fears. The image of the railroad more than any other new technology was associated with the disorientations of breaking frame: with the disorientation caused by increased travel speed, and fears of technology running out of control. In England, a national railroad system spread rapidly following the opening of the freight-carrying Stockton and Darlington line in 1825 and the inauguration of the Liverpool and Manchester passenger-carrying line in 1830. Americans initiated their first scheduled steam locomotive service in 1830 powered by the *Best Friend of Charleston*, built for the South Carolina Railroad, and in 1831 the *DeWitt Clinton* on the Mohawk and Hudson line, which ran from Albany to Schenectady. France in 1837 inaugurated its first common carrier with the Paris–Saint-Germain line, and railroads soon spread rapidly throughout continental Europe.

Those who kept diaries and notebooks recording their experiences riding on the earliest railroads revealed their feelings of vulnerability as they encountered the hazards of train travel. During his visit to America in 1837–1838, the British writer Captain Frederick Marryat, whose novels were among the most widely read in England and America in the 1830s, wrote of the dangers posed by sparks flying from the locomotive: "The locomotive was of great power, and as it snorted along with a train of carriages . . . it threw out such showers of fire, that we were constantly in danger of conflagration." He added: "The ladies, assisted by the gentlemen, were constantly employed in putting out the sparks which settled on their clothes."[24]

During his railroad ride from Boston to Providence on July 22, 1835, the American Samuel Breck, writing in his diary, noted that worries about dangers of travel seemed to take precedence over gallantry. When the train superintendent told male riders to "make room for the ladies!" by jumping up to ride on top of the train carriages, one passenger resisted, saying "I'm afraid of the bridge knocking my brains out." The passengers, Breck added, "made one excuse and some another."[25] In his diary entry of December 1839, he took a peevish view of the railroad, complaining that the "modern fashion" to "push on, keep moving," to "dash away, and annihilate space," had taken precedence over "comfort, security or pleasure." Railroad riding, he noted sardonically, would be the perfect form of travel "if one could stop when one wanted, and if one were not locked up in a box with fifty or sixty tobacco-chewers; and the engine and fire did not burn holes in one's clothes; and the springs and hinges didn't make such a racket; and the smell of the smoke, and of the oil and of the chimney did not poison one; and if one could see the country, and were not in danger of being blown sky high or knocked off the rails." He concluded that he would cling to "the old-fashioned way of five or six miles an hour, with one's own horses and carriage," for this would give him the liberty of seeing the country and being the "master of one's movements."[26] It was this feeling of mastery or control that seemed threatened by the railroad, a machine that, while bringing excitement at the annihilation of time and space, also brought a longing for the ability to regain mastery, to tame technology.

For nineteenth-century travelers, the most exciting and disorienting feature of the railroad was its speed. Railway managers favored slower speeds because they lessened the danger of accidents and kept down operating costs; it was the public that was exerting pressure for faster travel. As one manager complained in 1831, "The expectations of the public have been so much excited in reference to rapid travelling (and that must be by locomotive steam power) that they will not be satisfied with moderate speed, say 10 to 11 miles per hour, they must have 15 as a regular business."[27] By modern standards, of course, the speed of early railroads was not breathtaking. Although the American locomotive the *Experiment* reportedly reached more than sixty miles per hour on the Mohawk and Hudson line in 1832, New York passenger railroads averaged only fifteen miles per hour in 1840 and twenty-four miles per hour in the mid-1850s, an average that included the speed of express trains.[28]

The issue of railroad speed remarkably revealed the nineteenth century's optimism and anxieties about the new technology. The public's hunger for higher velocity coexisted with fears that faster travel would cause grave physiological and psychological harm.[29] Though many railroad passengers experienced a sensation of flying, as reported by Fanny Kemble, there was often a short distance between this feeling of flying and a lurking anxiety that the railroad was rushing out of human control.

Images of speeding railroads reverberated in the nineteenth-century consciousness and were made mythic by Thoreau who, hearing the sounds of a train in the quietude of Walden Pond, thought of Atropos, the emblem of unswerving fate.[30] Indeed, the railroad in literature and art became the embodiment of awesome power and lost human control—a harbinger of the next century's haunting fears of unpredictable, recalcitrant technologies.

Charles Dickens, who traveled by railroad during his visit to the United States in 1842 (and who would later experience a serious railroad accident at Staplehurst in England in 1865 involving a derailment in which his was the only first-class carriage not to fall over the side of a bridge), reported his ride on an American railroad in *American Notes*, where he presented a vision of American technology as speeding recklessly, little concerned with human safety. The train "rushes across the turnpike road, where there is no gate, no policeman, no signal: nothing but a rough wooden arch, on which is painted, 'WHEN THE BELL RINGS, LOOK OUT FOR THE LOCOMOTIVE.' On it whirls headlong . . . on, on, on—tears the mad dragon of an engine with its train of cars, . . . screeching, hissing, yelling, panting."[31]

Dickens's view of a relentlessly speeding railroad is vividly presented in *Dombey and Son* (1848), where the dazed Carker meets his nemesis as he is killed by an oncoming train, but the writer's most chilling portrayal of railroad dangers is seen in his short story "The

Signalman," which tells a tale of a ghostly figure that haunts a railroad signalman and warns him of impending accidents. The story's narrator, a signalman, descends to a railroad bed situated in a precipitous cutting of rock, in a "solitary and dismal place," and his descent is suggestive of one into hell. The signalman reports seeing the specter of another signalman announcing danger, and is later killed as he walks with his back to a speeding train that cannot stop in time. The signalman who is killed in Dickens's story represents mechanized man, a figure who works with "precision and exactness" yet cannot himself avoid being destroyed by technology.[32]

In the work of artists such as the great nineteenth-century painter J. M. W. Turner, the speeding railroad became an insistent technological presence, moving inexorably toward the viewer. Turner's vision was one of a somewhat mysterious and formidable machine, moving with an inevitable, but not fearsome, momentum. But the work of satiric artists more clearly suggested covert fears of technology eluding human control. Their graphics often echoed public alarm while representing playful ways to tame the new technologies. One of the most telling spoofs of mechanical speed was *The Flight of Intellect; Portrait of Mr Golightly, experimenting on Mess Quick & Speed's new patent high pressure Steam Riding Rocket* (c. 1830, fig. 15). This lithograph by an anonymous artist shows a frock-coated gentleman, "Mr Golightly," in striped pants riding a rocket-shaped steam vehicle through the air, a spoof on George Stephenson's locomotive the *Rocket*, which was one of eight locomotives participating in the Rainhill speed trials held in 1829 to select a locomotive for the Liverpool and Manchester Railroad.

At the Rainhill trials, the *Rocket* admirably outperformed all others, with speeds up to twenty-four miles per hour, but its victory was marred at the opening ceremony of the Liverpool and Manchester Railway in 1830 when the locomotive accidentally ran over William Huskisson, a member of Parliament from Liverpool who died from injuries a few hours later. In the artist's satiric view, the "March of Intellect" becomes *The Flight of Intellect*, and Mr. Golightly's figure, which straddles the steam rocket like a horseback rider, suggests an effort to master the machine rather than be overpowered by it.

The image of Mr. Golightly astride his rocket engine was echoed in very similar prints by other artists of the period, as seen in a Viennese print, *New Travel on a Steam Engine* (1846), which shows a man, his cape flying and his foot in stirrups, riding through the air on a steam rocket over the stately buildings far below. An anonymous

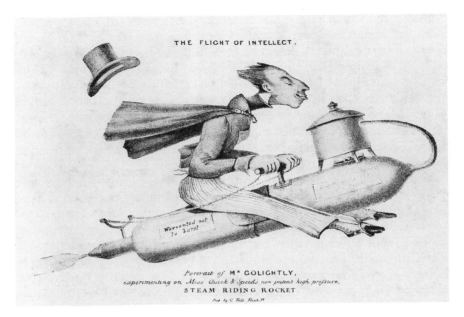

15 *The Flight of Intellect; Portrait of Mr Golightly, experimenting on Mess Quick & Speed's new patent high pressure Steam Riding Rocket* (c. 1830). Trustees of the Science Museum (London).

British artist of the same period, in his print *Elopement Extraordinary*, even envisioned a flying couple astride their steam rocket embarking on their "Matrimonial Excursion to the Moon, on the New Aerial Machine" (n.d., fig. 16).

The image of such a happy couple on their aerial flight would be echoed a hundred years later in American automobile advertising of the 1950s, with the image of a smiling family flying through the air astride their "Rocket Oldsmobile." In contrast to these benign views of technology safely under human control, the straddled rocket imagery also had a mythic power to suggest technology out of control, as American film director Stanley Kubrick recognized in his film *Dr. Strangelove* (1964), where air force Major T. J. "King" Kong straddled a rocket-shaped nuclear bomb as he rode like a maniacal cowboy to an explosive global catastrophe.

The railroad, in the hands of nineteenth-century caricaturists and illustrators, frequently came to represent an amalgam of anxieties about new steam technologies. Fears of excessive speed and railroad accidents remained all-too-real nineteenth-century concerns, starting with the earliest catastrophes caused by fractured wheels, broken axles, exploding boilers, signal and braking errors, and mistakes by both passengers and train crews. Boilers on early railroads might be only infrequently examined, and safety valves could be tampered with by engineers needing more steam pressure. In the United States, the *Best Friend of Charleston*, running on wooden wheels and iron tires, blew up when a fireman held down the safety valve; the exploding boiler was thrown twenty feet into the air, and the fireman died of injuries several days later. On English and American railroads after 1870, the number of boiler explosions declined steadily due to less tampering, more frequent and thorough inspections, and a revolution in boiler making as strong Bessemer steel replaced iron plates and fewer seams were used.[33]

Accidents caused by human errors often reflected the disorienting impact of new technologies as passengers and crews found themselves struggling to adapt to new experiences. Engineers blinded by wind, locomotive sparks, and steam often could not see railroad sig-

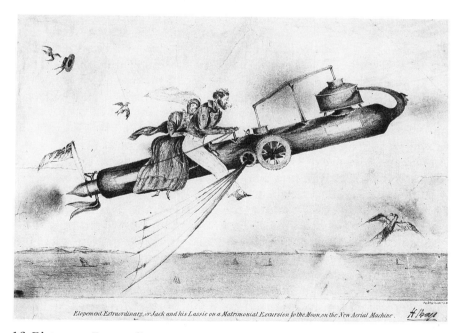

16 *Elopement Extraordinary, or Jack and his Lassie on a Matrimonial Excursion to the Moon, on the New Aerial Machine* (n.d.). Trustees of the Science Museum (London).

nals. Railroad passengers modeled their behavior on previous experiences riding stagecoaches, but their actions were often dangerously inappropriate. Passengers, accustomed to riding on the high, outside seat of stagecoaches, sometimes rode on top of railroad cars, unmindful of the possibility of being crushed against arches and bridges. Unable to gauge the new speeds, they were hurt as they jumped from moving trains without waiting for the train to come to a stop at the station. Early railroads lacked station platforms, and passengers were allowed to roam freely with the often disastrous consequence of being hit by a train coming from an unexpected direction.

British science professor Dionysius Lardner argued that, statistically, the risk to a passenger of being involved in a railroad accident was small, yet he held that during the period 1850–1851, more than half of accidents involving passenger fatalities were due to the passengers' imprudence or errors. Though some contemporaries viewed Lardner's knowledge of science as woefully inadequate, what remains revealing in his writings on the railroad are his "rules for travellers" meant to help naive passengers reduce their risk of accidents. Riders are advised, "Never attempt to get into or out of a railway carriage when it is moving, no matter how slowly," for "it is a peculiarity of railway locomotion, that the speed, when not very rapid, always appears to the unpractised passenger much less than it is. A railway train moving at the rate of a fast stage-coach seems to go scarcely as fast as a person might walk" due to "the extreme smoothness of the motion." Lardner's rules also warn of other common misjudgments by inexperienced railroad riders: "Never get out of the wrong side of a railway carriage" (in the path of oncoming trains), and "beware of yielding to the sudden impulse to spring from the carriage to recover your hat which has blown off."[34]

To help prevent accidents, nineteenth-century railroad companies standardized travel regulations: the Liverpool and Manchester Railroad in 1831 set speed limits of not more than seventeen miles per hour and printed regulations for signal and whistle systems in 1840. American inventor George Westinghouse greatly improved railroad safety with the patent for air brakes in 1869. Yet in spite of improved safety measures, accidents continued to occur with grim regularity and were assiduously reported in nineteenth-century newspapers and periodicals in the United States and Europe, though not always with great accuracy.

Although the subject of railroad accidents was rarely treated by nineteenth-century painters, journalistic reports were often accom-

17 "The Wreck of a Moody and Sankey Excursion Train." *Frank Leslie's Illustrated Newspaper*, February 2, 1878.

panied by artists' drawings or wood engravings that magnified the horrors, showing human casualties lying in rows near derailed trains. Publishers of American weeklies during the second half of the nineteenth century, as in Europe, capitalized on the melodrama of railroad accidents, magnifying the terrors of railroad travel. By providing detailed if somewhat sensationalized illustrations of accidents, newspaper artists helped establish the dangers of railroad travel as historic fact in the public imagination. In an age before halftone photographs largely replaced wood engravings in periodicals, railroad accidents were, in effect, made real through artists' documentation (fig. 17).

But artists, while creating what appeared to be documentary views, were also often engaged, in a broader sense, in helping to tame the new technology and soften public anxieties by presenting idealized images of railroad travel. In 1872, for example, *Harper's Weekly* engaged in mythmaking by telling a generic tale of a railroad disaster averted by the vaunted heroism of America's train crews. The story's illustration, titled "Danger Ahead," shows the train's engineer and fireman reacting to an unspecified dreadful occurrence by braking the train. As described in the article, the engineer is seen using one hand to pull the lever that stops the steam while the other hand pulls the lever that stops the action of the drive wheels. The fireman, meanwhile, gives the "down brakes!" signal and seizes the wheel of the engine-brake, bringing the train to a safe halt.

The *Harper's* story, clearly intended to instill confidence in the safety of railroad travel, lavishly praises the "hero of the track." Even so, it also offers an unintended countermessage: the sense of safety is undercut by the story's evangelical tone giving thanks for a dreadful danger barely avoided. After the train's safe stop, "a voice goes up to Heaven from within the cars in praise for the great deliverance."[35]

American chromolithographs, which were being produced after 1850 by such leading publishers as Currier and Ives, provided panoramic images of railroads which further idealized the presence of technology in the American landscape and helped promote the new express trains and luxury "palace" cars. These commercial prints were often issued as works of art, but some versions also included added railroad company advertising text. The Currier and Ives print *An American Railway Scene, at Hornellsville, Erie Railway* (1876), for example, which dramatically illustrates several cars including a Pullman parlor car and sleeping coaches, bears advertising in some versions promoting tickets for the Erie Railway.[36] In advertisements for their own services, lithographers courted business from railroad builders by emphasizing their accuracy in drafting details of railroad design—making the prints suitable not only as advertisements to be hung on the walls of railroad company offices but also as documentation for master mechanics.[37]

The Currier and Ives railroad prints, issued in black and white and sometimes color-printed versions, were largely celebrations of American railroads and epitomized the country's pride in its technological achievements. In the year before the completion of the transcontinental railroad, the prints also helped promote the idea of the railroad as a potent force in forging national unity, bearing titles such as *Across the Continent: "Westward the Course of Empire Takes Its Way"* (1868).

During the years 1830–1850, the early period of British railroad development, commercially sponsored prints by artist John Cooke Bourne and those of Thomas Bury with his stately, well-dressed passengers also became a testimony to railroad safety, providing apparent proof, as it were, of the civilized nature of railroad travel. British painters, too, helped soften the impact of the emergent railroads. When Isambard Kingdom Brunel, the talented builder of Britain's Great Western Railway, sought to quell public anxieties about the safety of the Thames Tunnel, a project first conceived of by his father, Marc Isambard Brunel, he held a banquet in the tunnel for 150 people—an event captured by an anonymous artist whose oil painting records the scene complete with a long banquet table covered with a white cloth, seated frock-coated guests, and candles burning in graceful, curvilinear brass candelabra. The oil painting, in effect, helped legitimize the tunnel project by lending an air of dignity, civility, and elegance to the event.

While painters and commercial printmakers in the service of the railroad companies helped give the railroad an aura of familiarity and safety, British and other European artists also used humor as a strategy to ease anxieties about the railroad, though their comic views tended to censure rather than civilize the machines. But American illustrators, often adept at capturing the dangerous aspects of railroad accidents, rarely presented satiric views of the more serious hazards and traumas of travel by rail. When they did occasionally spoof railroads, their images generally were not of lethal dangers but instead poked gentle fun at the drolleries of travel. The crudely drawn sketches called "Humors of Railroad Travel," published in *Harper's Weekly* in 1873, presented such scenes as passengers inconsiderately propping their feet on the seat ahead, a passenger trying to make his way through the aisles of a lurching train, and a newly married romantic pair captioned "new car coupling on the honeymoon route."[38]

Currier and Ives issued a small number of railroad cartoons published separately as prints, several of which were included in its "Darktown" series of the 1880s, which engaged in racial if not racist comedy about American blacks that made gently humorous appraisals of railroad hazards and misadventures. *Off His Nuts* (1886) shows a wide-eyed, frightened man colliding with a train while riding his high-wheeled bicycle. Terrified, he wraps his body around the train's smokestack and announces, in stereotypical dialect, "Gracious Massy, I'se Struck de Comet!" In a print that plays on sanctions against interracial romance while taking a comic swipe at railway travel, *A Kiss in the Dark* (1881) presents two panels, one portraying a prank-playing conductor who announces that the train will be traveling through a dark tunnel for one-half hour. An elderly gentleman sits reading while a seated young white man glances back at a young white woman, who sits next to a black nanny holding a white child. The second panel presents the aftermath of the conductor's prank: the train actually emerges from the tunnel in three minutes, and the light of day reveals the elderly gentleman now drinking whiskey from a flask and the young man embracing the black nanny, who had apparently switched seats with the young woman.[39]

Although American artists and their publishers could hardly avoid recognizing the railroad's potential dangers, they were less apt than their European contemporaries to humorously belittle travel by rail. They lived, after all, in a society that popularly considered the new system of transport a boon rather than an economic or social threat. As cultural historians have long noted, nineteenth-century Americans tended to view the railroad as perhaps dangerous but also as an essential improvement to the existing transportation systems of canals, steamboats, and stagecoach travel over largely unsurfaced roads. American factory owners, farmers, and merchants, impatient with boats and wagons, welcomed fast and dependable transportation. The American continent itself was vast, and distances needed to be bridged; the railroad became a valued means of forging national unity and fulfilling the nation's wish to transform the wilderness and open up new western territories.[40]

In Europe, however, comic views of the railroad were popular fare, in part because artists were sensitive not only to railroad hazards but also to the humorous ways the machines were unsettling European social conventions. Experimental British steam carriages and the newly developed railroads became ready satiric targets beginning in the late 1820s, a time that coincided with the tail end of England's most prolific period of satiric printmaking.

Comic satires of the railroad were also popular in France, with its established tradition of social and political satire in place and ready for the topical subject of new technologies. Lithographs were a potent medium for satire in the French political weekly *La Caricature* begun by the caricaturist Charles Philipon in 1830. Two years later, Philipon introduced *Le Charivari*, which became an important vehicle for the brilliant satires of Honoré Daumier.

French artists, notably Daumier, were particularly skeptical of the railroad, reflecting perhaps the great anxiety produced by France's first major train catastrophe, which occurred in 1842 on the Paris–Versailles line. The accident, in which a train en route to Paris broke a locomotive axle and killed fifty-seven passengers, who burned to death when colliding carriages caught fire, led almost immediately to the enactment of the French national railroad regulatory laws in 1843.[41]

Nineteenth-century European satirists not only mirrored public anxieties about the new autonomous technologies which raced on, indifferent to human safety, but also helped tame technology by allowing viewers to confront conscious as well as buried fears of being ridiculously toppled or, worse, of being mangled, burned, and mauled by errant machines. Caricaturists delighted in sketching the pitfalls awaiting those passengers who deserted traditional modes of travel for new machines, and their comic images often underscored the *Punch* magazine credo: "A railway is long, but life is short—and generally, the longer the railroad, the shorter your life."

HONORÉ DAUMIER

The most incisive and mordantly comic views of railroad travel came from the hand of Honoré Daumier, whose fluid, energized drawings and highly observant eye made him the dominant European caricaturist of the nineteenth century.[42] Born in Marseilles in 1808, Daumier moved with his family to Paris in 1816, where he mastered the new medium of lithography and began contributing to *La Caricature*, the weekly founded in 1831 engaged in satirizing the Revolution of 1830. After being arrested and briefly imprisoned for his satiric drawing of French "citizen-king" Louis Philippe, Daumier in 1832 began regularly contributing to the journals *Le Charivari* and *Le Boulevard*, prompting Baudelaire to praise the artist for his remarkable ability to capture "every little stupidity, every little pride, every enthusiasm, every despair of the bourgeoisie."[43] During his lifetime, the prolific satirist produced over four thousand lithographs on human foibles and the absurdities of French society.

Daumier's caricatures of the railroad first appeared in 1843 with the publication of *Les Chemins de Fer*, a series of lithographs printed in *Le Charivari*, followed by his series of ten lithographs *Physiognomies des Chemins de Fer* (1852), a second *Chemins de Fer* series in 1855, and others in 1857–1858, 1862, and 1864. In his views of the railroad, the artist was often preoccupied with two recurrent nineteenth-century themes: the discomfiture of people engaging with technology, and technology rushing out of control. The railroad became a ready target for Daumier, who pictured passengers being jostled about, their physical and mental equilibrium thoroughly shaken. In the 1840s, when France, England, and the United States were rapidly building national railroad networks, trains (which averaged about 20–30 mph) seemed the embodiment of runaway technology, of machines going too fast to take heed of mere mortals. His satires of speeding machines became a means of taming technology, a means of easing the pain, if not the pace, of technological change.

Daumier's railroad art represented some of the century's most pointed and biting views of the machine's impact. His lithographs, with their incisive, elegant, comic exaggerations, were particularly adept at capturing the indignities suffered by train travelers. In a print from his *Chemins de Fer* series of 1843, passengers jammed into a string of open-air third-class train cars ride with their heads thrown back and their arms toward the sky in alarm and despair (fig. 18).

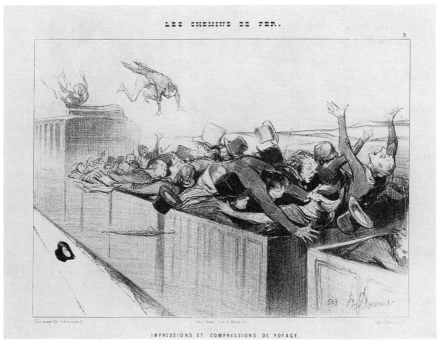

18 Honoré Daumier, "Impressions et Compressions de Voyage," *Les Chemins de Fer* (1st ser.). Lithograph. *Le Charivari*, July 25, 1843. The Metropolitan Museum of Art, New York, A. Hyatt Mayor Purchase Fund, Marjorie Phelps Starr Bequest, 1989.

19 Honoré Daumier, "Conducteur!" *Les Chemins de Fer* (1st ser.). Lithograph. *Le Charivari*, October 30, 1843. Gift of William A. Sargent, courtesy, Museum of Fine Arts, Boston.

One man's body flies through the air, while another passenger complains, "Mercy, we are all lost." A nearby passenger offers scant comfort: "It's simply that the train is starting up again . . . from the moment the train goes forward, the passengers go backward, everyone knows that." Daumier's lithographs present the train as an inflexible technology fixed on its own course, a technology indifferent to human needs. As the train emerges from a tunnel, a sick passenger calls out that he has a stomachache, but the trainman refuses to stop: "Impossible. It's against regulations. But in two and a quarter hours, we will be in Orléans" (fig. 19).

Daumier's caricatures presented all-too-real discomforts and hazards of railroad travel, and they reflected the artist's skepticism of a system of transport that tended to undermine human dignity, transforming passengers into timid, comic, foolish figures. In his satires, images of human vulnerability abound, as seen in his views of third-class passengers, who in early French trains rode standing in seatless, open boxes-on-wheels where passengers were often jammed together and were occasionally drenched by rain while riding hatless in their roofless carriages.

Daumier and other European artists were sensitive to class differences in passenger coaches. Europeans in the 1840s admired the fact that Americans had no distinction between first- and second-class coaches, and Americans persisted in maintaining this mythology about single-class railroad travel, though there were indeed instances of second-class cars, roof seats for economy class, separate cars for men and women, as well as separate seats or cars for black Americans and, after midcentury, better-appointed parlor cars and luxurious sleepers designed for exclusivity.[44] Daumier, however, acknowledged the reality of class differences and saw the comic possibilities of (sympathetically) satirizing the perils of third-class railroad travel and the vulnerability of even first-class carriages.

But there was more than passenger discomfort reflected in Daumier's lithographs. They also revealed the most profound and recurrent fears associated with railroad travel and other technologies as well: the fear of losing limbs, of being blown apart by mechanical explosions, of being unexpectedly thrown off or unable to stop a speeding vehicle. One of Daumier's most overt references to the nightmarish threat of dismemberment was in his lithograph "The Entrance to the Large Tunnel" (1843), which pictures several passengers in top hats riding on the roof of a train carriage, unaware of imminent danger, while another rider anxiously warns, "We are going to enter into a huge tunnel. . . . I urge you, don't move during the trip—there hasn't been a trip without someone losing an arm, a leg or a nose. And you know it's impossible for the Administration to retrieve them in a dark subway two leagues long." It is little wonder that an anxious elderly couple in another print see the railroad as carrying diabolic passengers who "sit in a train of hell with their devil's boiler."

Daumier's characteristic view was that of a technology which endangers human safety and sense of self, a speeding machine moving inexorably according to its own imperatives. His caricatures crystallized many of the century's most troubling fears of mechanization and reflected his uncanny perception of the profoundly serious embedded in the seemingly silly. Daumier's art caricatured what was to become one of the central criticisms of both technology and industrialization: the demand for technological efficiency was resulting in a ruthless depersonalization which turned human beings into commodities or mere objects, illustrating John Ruskin's lament in *The Seven Lamps of Architecture* that "the whole system of railroad travelling is addressed to people who, being in a hurry, are therefore, for the time being, miserable. . . . It transmutes a man from a traveller into a living parcel."[45]

Daumier's lithograph "Thirty Seconds at the Station" satirizes the railroad system's cavalier attitude toward passengers. During a quick station stop, a wrapped package and a male passenger with hat flying off his head are tossed off the roof of a railway car as one railroad worker calls out to another below: "Hold on, Joseph! This is all that I have to send you: three packages and a traveller. Take special care of the packages!" With his gift for observation, Daumier captures the passenger's predicament shown in his startled face, raised hands, and lifted knees as he falls through the air, while, as is typical in Daumier's caricatures, bystanders watch the event unperturbed.

Daumier's images of the railroad not only captured the shock of breaking frame but also again manifested the public's fear of violent death by machine. His most brilliant lithographs in his *Chemins de Fer* series achieve a comic tension by juxtaposing the casual with the catastrophic. In his art, the devastating is cloaked by banality, the moment of intense anguish viewed with absurd philosophical detachment. Anxious passengers in one lithograph ride in a horse-drawn stagecoach and complain to the driver, "Your coach isn't advancing!" as they all sit on a seat facing the viewer, while in the background, unseen by them, are a dangling train derailed over a viaduct, an upset locomotive, and human figures violently thrown high into the air. In another print, a stoic coach driver who sits next to a sleeping passenger comments with nonchalant irony, "We have time—I'm not driving to the moon—if you want to get there, you would have to take the train—I only go to Bougival."

Daumier continually achieved a comic deflation of the railroad by contrasting the calm and the hysterical, the anxious and the unalarmed. His linking of a rational response to emotionally charged imagery brilliantly tapped deeply disturbing anxieties about technology. In a lithograph of 1843, a man with his hands casually in his pocket surveys a stationary locomotive in a railyard, having just queried a nearby railroad worker. The trainman replies, "You ask me if my locomotive is good? Of course. . . . It is capable of leading us to America." The gentleman responds, alluding to the afterlife: "You mean to the other world? In that case, I won't go with you." Daumier's skillfully drawn station scene has a calm, classically balanced stability, forming an ironic backdrop for talk of death and disaster.

Though the printed captions for Daumier's drawings were often written by his publishers, it was the artist's ability to economically capture the subtitles of human gesture that gave his lithographs their arresting quality. He effectively caught the postures of railroad travelers who confronted fears of train accidents with an attitude of rationality and nonchalance. Two men sitting in a railroad car, for example, briskly discuss railroad accidents while a woman passenger looking ill leans sickly against the carriage window. "It seems to me that we are going to derail," one man comments anxiously, while the other replies with a look of absurd complacency: "You are afraid . . . not me! I'm insured for 100,000 francs. I wouldn't have anything against an accident in order to be paid an indemnity by my company."

Nineteenth-century European railroad travelers had to contend not only with the possibility of accidents but also with the lurking threat of assault by fellow travelers. Passengers who rode in the stagecoach-like first-class carriages often felt isolated and vulnerable to attack, particularly since poor communications systems made it difficult to get attention from a train worker (it was not until the 1860s that train cords in Europe were run outside carriage windows and, when pulled, opened a steam whistle connected to the engine). Daumier's lithographs satirized passengers' fears of assaults and their suspicious paranoia. In a lithograph of 1864, a passenger standing at a ticket window is informed, "Each first-class passenger has the right to have a ticket and a pair of pistols," while in another view, two men seated in a first-class coach eye each other suspiciously with pistols in their hands.

Daumier's work highlighted passengers' anxieties about the unfamiliar, and his comedy only thinly masked more profound fears, particularly the feeling of railroad travelers that they had no personal control over the new technology. In a revealing lithograph, two men converse in a coach while a third looks on anxiously, thinking, "They seem to conspire. I am angry with myself for being in this railroad—with all this—one doesn't have a way to stop the train." The artist took a more lighthearted view of third-class travel, which in his view brought discomfort but little danger. Illustrating class disparities, his lithograph of 1864 presents the smiling figure of Monsieur Prudhomme—the nineteenth-century French archetype of the com-

20 Honoré Daumier, *The Third-Class Carriage* (c. 1863–1865). Oil on canvas. The Metropolitan Museum of Art, New York, Bequest of Mrs. H. O. Havemeyer, 1929, The H. O. Havemeyer Collection.

placent bourgeoisie. Seated among a crowd of third-class travelers, Prudhomme comments happily, "Long live the third-class carriages. One may be asphyxiated, but never assassinated!"

From 1856 to 1865, Daumier not only produced satiric lithographs but also paintings and drawings on the subject of railroad travel. These images of passengers in first-, second-, and third-class carriages were not so much comic views of the perils of railroad travel as closely observed character studies. Among the paintings of passengers, one of the best known is *The Third-Class Carriage* (1863–1865, fig. 20), a work that attains a degree of warmth and intimacy not seen in his comic lithographs. As he focuses on the benign, gentle faces of a hooded elderly woman, a sleeping young boy, and a mother who cradles her infant child, Daumier here ceases to view the railroad as an adversary; it has become domesticated, peaceful enough for mothering.

His crayon and watercolor painting *First-Class Carriage* is a study of two male and two female passengers sitting with content absorption in their own thoughts as one reads a book and another gazes out the window. The passengers are in a state of relaxed equilibrium; the artist deftly uses strong contrasts of light and shadow to outline their features and lend drama to an otherwise peaceful scene. Markedly absent are the exaggerated, grotesque features of anxious passengers in Daumier's caricatures, where faces are often contorted by bewilderment and alarm.

The passengers in Daumier's paintings, as well as passengers in Victorian-era British paintings and in illustrations of luxury train interiors in American weeklies, are often seen as relaxed riders comfortably adjusted to railroad travel—a reflection in part of railroad manufacturers' increasing efforts after midcentury to create a greater sense of comfort. Travelers in early coach cars during the 1830s sat on hard wooden seats, but by the 1840s American coach seats were upholstered, and during the 1850s increasing attention to passenger comfort on the more luxurious cars brought a boom in American patent railroad furniture, including reclining seats.

Although Americans ostensibly had no class differences in coach types, a number of carriage designs intended for luxurious, first-class travel were introduced during the 1860s, including parlor cars or day coaches, dining cars, and sleeping cars ("palace cars") fitted with elegant, wood-paneled interiors with velvet wall hangings and silver appointments. These sumptuous improvements were artfully detailed by illustrators whose work was commissioned for railroad advertisements and whose commercialized views were echoed in illustrations that appeared in *Harper's* and other American weeklies.[46] British and French railroads had similar luxurious accommodations including, in England, the Stockton and Darlington line's sumptuous smoking cars of 1863, with thick tapestry curtains and interiors much like American luxury cars of the period.[47] British and French second- and third-class travel also benefited from improvements in comfort, including upholstered seats and backs.[48]

Reflecting this increased sense of comfort, European and American artists presented somewhat sentimentalized views of railroad carriages, particularly the British artists after midcentury. While Daumier and Britain's Sir John Tenniel in his cartoons for *Punch* continued to satirize the railroad after midcentury (Tenniel, through 1890, was publishing macabre reminders of railroad accidents), other artists were simultaneously presenting reassuring images of railroad travel which drew closer to the subject and portrayed more intimate views of human relationships.

Although passengers were still shown riding in dreary conditions,

21 Abraham Solomon, *The Return* (1855). Oil on canvas. Courtesy of The National Railway Museum, York, England.

or being violently hurt by collisions, derailments, and runaway trains, they were also depicted as being comfortably nested in their railroad carriages and having a more integral connection with technology. Through this intimate focus, artists helped tame lurking anxieties about train travel by treating railroads as cozy, familiar, cushioned settings for moments of warmth, romance, and family feeling.[49]

In paintings as well as in newspaper and magazine illustrations after 1850, the interiors of railroad carriages became a new pictorial arena for intimate vignettes, a miniature stage on which were enacted small human dramas. Framed by the close boundaries of the train carriage walls, passengers are shown sadly parting, conversing, and even courting. While passengers in Daumier's *Chemins de Fer* series of the 1840s are anxious travelers battling fears of the unknown, railroad travelers after midcentury were shown feeling secure enough to engage in romantic flirtations, as seen in the painting *The Return* (1855, fig. 21) by British artist Abraham Solomon, who produced several versions of first-class railroad travel.[50]

Sitting on richly upholstered stuffed-leather seats, a father and a young man dressed in a British naval uniform talk intently while the daughter, sitting at the far end, gazes fixedly at the young man (in an earlier version that caused controversy, the father sleeps while the young couple, sitting in adjacent seats, flirt with each other). In Solomon's painting, the view of the landscape seen obliquely through the train window becomes in itself a framed painting—a picture on the wall in this domesticated, homey interior.[51]

Some of the most telling examples of this theme of integration and adaptation can be seen in the paintings of the French Impressionists, whose divisionist style, paradoxically, conveys images of the railroad's integration into the landscape. In Claude Monet's *The Train in the Countryside* (1870), the familiarization of the railroad in art is largely complete. Here the railroad is known less by its shape than its sign: seen from afar, a train is partly obscured by a row of trees, its presence indicated largely by trailing puffs of smoke and steam. Smoke becomes a sign of the partially hidden train's existence. The artist is not engaged in documenting the train as a novelty or in reflecting on and reducing travel anxieties through satire. The railroad is no longer camouflaged by the landscape but rather is attuned to the terrain—a presence that need only be indicated, and acknowledged, in pictorial code. During the next century, artists who were concerned about the fragmenting impact of the machine would turn their attention to other, more threatening technologies that posed new challenges to the human frame.

Chapter 2

Art, Technology, and the Human Image

As seen in their turbulent images of exploding steam engines and runaway trains, nineteenth-century artists were sensitive to the traumatized emotions that often accompanied dramatic changes in technology during the period. In an era of experimental steam coaches and expanding railroad systems, travelers were confronted with intensified perceptions of their own vulnerability, and the fragile tentativeness of the human frame. But nineteenth-century art also reflected other, perhaps more subliminal changes in human self-conception that accompanied rapid mechanization. Those who lived during this period of disjunctive technological change felt not only rude psychological jolts but also dramatic shifts in their own sense of stature, size, and scale.

Images of massive steam engines tended to suggest an expansive pride in Promethean technological achievements, yet caricatures of the period tended to reduce human stature by belittling the century's mania for mechanical inventions. In paintings of industry, even dwarfed human figures were often ennobled in their labors; yet, as one of the unresolved tensions of the era, satiric artists also saw both a loss of dignity and a source of hilarity in people experiencing the hazards and humiliations of adapting to new machines. These twin views of human stature in the context of technology made vividly clear the extent to which the new technologies—through their often gigantic size, through their powerful and sometimes hazardous nature, through their capacity to wonderfully enhance human abilities—were swiftly redefining the parameters of human identity.

TECHNOLOGICAL GIGANTISM AND HUMAN SCALE

In a century that relished technological gigantism, one of the recurring motifs in nineteenth-century art was the image of minute human figures standing in the context of towering emblems of technology. In early pictorial views of British industry, artists frequently exaggerated the size of industrial structures, an exaggeration that inevitably created a peculiar tension between the structures and the figures in the work. British artist George Robertson, an eighteenth-century landscape painter and drawing master whose oil paintings were redone as engravings by other artists, emphasized this disparity in size and scale in his pictorial views of the iron industry in England. His engraving *The Inside of a Smelting-House at Broseley* (1788, fig. 22) shows a blast furnace at night illuminating the bent backs of ironworkers pouring molten iron into oblong troughs. In the engraving, the

22 Wilson Lowry after George Robertson, *The Inside of a Smelting House at Broseley, Shropshire* (1788). Line engraving. Ironbridge Gorge Museum Trust, Elton Collection.

foundry is transformed into a huge amphitheater, with its height emphasized by the soaring timbers of a crane and roof supports that loom high over the laborers.

In Robertson's art, and more clearly in later nineteenth-century works, the vast gulf between the dwarfed human figures and gargantuan structures of industry was not intended to belittle people but to salute their technological abilities. There is a strong suggestion of pride and wonder directed toward industrial works; in Robertson's work, there are overtones of the awesome in the nighttime visions of the technological sublime.

The disparity in size between man and machine could also signal a more pessimistic view and sense of despair about the potentially deadly dominance of machine technologies over human beings—a view in which human figures were ultimately devalued. Gigantic images of machines suggested pride in the heightened possibility of dominating and controlling the forces of Mother Nature, yet sometimes the small size of human figures also suggested an implicit feeling of childlike helplessness as these gigantic technologies threatened to dwarf, dominate, if not destroy their human creators. Part of the fascination of Mary Shelley's gothic novel *Frankenstein* (1818) lay in its powerful evocation of the menacing, artificially created monster whose eight-foot frame and "gigantic stature" became a source of wonder and terror. In his obsessive quest to re-create human life, Frankenstein deliberately made the figure huge because, as he claimed, normal-size humans had too many minute parts for him to duplicate. The potent force of Shelley's mythic tale arises from its ability to embody the century's dual views of technology: the expansive pride in human inventiveness and the terrifying fear of being overshadowed and even annihilated by one's own technological expertise.

An intriguing emblem of this dual perception of human stature was a little-known but remarkable invention—a sculptural and mechanical version of humanized technology—shown at London's Great Exhibition of the Industry of All Nations held at Joseph Paxton's Crystal Palace in 1851, where new developments in science and technology were proudly displayed for the six million visitors who flooded the glass-and-iron technological showcase. Amid the steam engines, threshing machines, ornamental ironwork, textile exhibits, and decorative arts, one exhibit received particular recognition, though it seemed eccentric and out of place: the steel "Expanding Figure of a Man" invented by Count de Dunin (fig. 23). Built of slid-

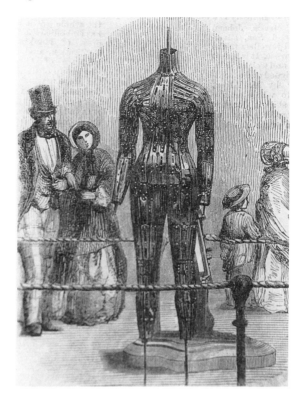

23 Count de Dunin, "Expanding Figure of a Man." *The Crystal Palace and Its Contents: An Illustrated Cyclopedia of the Great Exhibition of Industry of All Nations 1851* (London: Wm. Clark, 1852).

ing metal plates and tubes, Dunin's Expanding Man was designed to expand and contract, to change from lifesize proportions to gigantic stature. Its function was to reproduce the different proportions of the human figure, and according to the exhibition's official catalog, it could duplicate the deformities and peculiarities of any individual. The Expanding Man, which was considered suitable for an artist's studio, was designed for garment fittings and as a useful resource for army tailors who might have to construct uniforms for soldiers in distant colonies.

To one contemporary observer writing in an *Illustrated Cyclopedia* of the exhibition, Expanding Man represented "industry 'run mad.'" Noted the reviewer, "industry is one thing and caprice is another and very different thing."[1] Yet the Council of Chairmen of Juries at the exhibition apparently thought otherwise. After considering the Expanding Man, which consisted of over nine thousand largely concealed mechanical parts, the council awarded Dunin's work its "Council Medal" for "the extraordinary application of mechanics."

The council's award was not as quixotic as it might seem. The steel Expanding Man was an assertion, much like the innovative Crystal Palace building itself, of the century's great pride in its technological achievements. Intended as a utilitarian mechanism, the Expanding Man was included in the exhibition's section of "Philosophical, Musical, Horological, and Surgical" scientific instruments, yet the invention's human form transformed it into a work of art, a sculptural embodiment of nineteenth-century technological man. In their report, the exhibition's jurists referred to "this beautiful piece of mechanism."[2]

The Expanding Man, too, reflected the mythic, godlike stature often accorded to new technologies. The official catalog of the Great Exhibition likened the Expanding Man to a Greek god and to classical Greek art: the invention "admits of being expanded from the size of Apollo Belvedere to that of a colossal statue"—a linking of classicism and the machine which would become a recurrent motif for the next hundred years, informing the rhetoric of modernist architecture and design.[3]

The Expanding Man represented the nineteenth century's preoccupation with machine gigantism and also its dual vision of human scale. On the one hand, the human ego was swelled by the technological expertise manifested by large machines. As Lewis Mumford noted, gigantism was a coveted feature of nineteenth-century mechanization, since the engines run by steam power were considered most effective when built in large units. Gigantic size became a symbol of technological efficiency and progress.[4] On the other hand, many feared that machines and factories were becoming too massive and, as in the Frankenstein story, would turn people into dwarfed creatures in danger of mechanical domination.

The works of many nineteenth-century artists illuminate the impact of technological gigantism on human scale while revealing the century's complex response, its admiration and fear of massive technologies. In 1832, the British inventor and artist James Nasmyth introduced his giant steam hammer for use in his iron foundry near Manchester, and the hammer was later manufactured in varying sizes, with some over nineteen feet tall. However, when Nasmyth painted his *Steam Hammer at Work* in 1871, he greatly magnified the

24 James Nasmyth, *Steam Hammer in Full Work*. Engraving from a painting by James Nasmyth. *James Nasmyth, an Autobiography*, ed. Samuel Smiles (London: John Murray, 1883).

size of the hammer, which is shown towering over a line of ironworkers in a dark, fire-lit forge. The huge invention itself, in Nasmyth's view, is anthropomorphized, resembling a humanoid figure planted firmly on its spread iron legs (fig. 24). In the red glow of the foundry, the steam hammer with its exaggerated size is both awe-inspiring and powerfully fearsome, an apt embodiment of the century's ambiguously celebratory view of technology, which both empowers and overpowers its human creators.

Artists also celebrated gigantism in their engravings and lithographs of the industrial exhibitions that were so popular in nineteenth-century Europe and the United States. London's Great Exhibition of 1851 became the century's most famous and most documented display of technology, and it was particularly known for the unique exhibit hall itself. Graphic artists of the period strove to capture the pride and excitement generated by Joseph Paxton's mammoth Crystal Palace, which contained 800,000 square feet of exhibition space made light and luminous through panels of glass. Joseph Nash's chromolithograph of the palace's opening captured the drama of the event and the structure's monumentality, showing Prince Albert and Queen Victoria leading an inaugural procession while a throng of onlookers appear dwarfed by the soaring lines of the architectural frame of iron overhead (see fig. 48).

Illustrations in nineteenth-century weeklies also revealed the tendency of artists to exaggerate technological gigantism. Prints showing steam engines, Bessemer steel converters, and steamship boilers often rendered the machines as monumentally huge and human figures as relatively minute, emphasizing the difference in size. In a *Harper's Weekly* print of a Bessemer converter (1886; see fig. 26) by American artist Charles Graham, the usually ten- to fifteen-foot high converter has an exaggerated height as it looms over factory workers.

Many nineteenth-century machines were indeed massive, as periodical illustrations of the thirty-nine-foot tall, 680-ton Corliss engine featured at the Philadelphia Centennial Exhibition in 1876 proved to contemporaries. Though relatively few nineteenth-century painters considered new machinery worthy of artistic treatment, artists commissioned to draw newspaper and periodical illustrations provided much-needed visual information about technology. Their work not only reflected the century's perceptions about mechanization but also often rose beyond the merely documentary and emerged as elegant and dramatic depictions of the new technologies. A *Harper's Weekly* illustration of the Corliss engine based on a drawing by Theodore R. David accentuated the gargantuan size of the engine by placing small spectators in the foreground, thus emphasizing the massive wooden supports and ladders leading up to the engine (fig. 25). The exhibition visitors who surround the engine appear as well-dressed, stately figures, whose stature is ennobled rather than diminished by the machine.

There is no hint here of any fear about the machine's huge size—a fear that was, however, expressed by American novelist William Dean Howells. When viewing the Corliss engine at the Philadelphia Exhibition, Howells considered that the massive machine had the potential of crushing its engineer "past all resemblance of humanity"—a thought quickly subdued by expressions of pride in the exhibition's "glorious triumphs of skill and invention."[5]

Howells's comment is a testament to nineteenth-century America's ability to absorb anxieties about technology into a value system that saw mechanization as a means to national progress. The bone-crushing power of the massive Corliss engine becomes reformulated in Howells's view as the massive power of America's genius. As his

74   *Breaking Frame*

25 Corliss engine. Engraving. *Illustrated Catalogue of the Centennial Exhibition, Philadelphia 1876* (New York, 1876).

comments make evident, the monumental size of the engine was closely connected to its mythic meaning. Charles T. Porter, a noted nineteenth-century American engineer, made it clear in his *Engineering Reminiscences* that the Corliss engine's towering size may have had less to do with functional necessity than with heightening its emblematic stature. The engine design with ten-foot piston strokes and two cylinders each forty inches in diameter, and the great height of the engine's vertical frame, Porter complained, resulted in the engine's running at a slower speed. With less material and a horizontal frame, he argued, the engine would have run with greater economy and at a faster speed, albeit with a loss of its emblematic, monumental status.[6]

The small number of American artists who depicted technological subjects or drew illustrations of machinery with human figures standing nearby tended to represent the technologies as massive but unthreatening. A telling instance appeared in the *Harper's Weekly* article "A Revolution in Iron and Steel Manufacture" (1889), which described the Bessemer steel process that had only a month before been given an American patent. With characteristically effusive rhetoric, *Harper's* described the process as an achievement "which has, more

26 Charles Graham, "Making Bessemer Steel at Pittsburgh." *Harper's Weekly*, April 10, 1886. Photograph: Smithsonian Institution, Washington, D.C.

than any other modern invention, laid the broad basis for the material advancement of modern civilization." Charles Graham's accompanying illustration of a new Bessemer steel converter (a detail of his 1886 illustration, fig. 26) reveals the artist's awe of and admiration for the technology: in the factory darkness, a converter towers like a planetary sphere, spewing celestial sparks on three figures below, who continue about their factory work untroubled by the erupting mechanical presence that looms over them.[7]

Technological gigantism continued to be a popular feature in nineteenth-century American weeklies, which often displayed cover illustrations in which the huge cylindrical forms of machines were juxtaposed next to human figures for a sense of scale. But here again, there was no loss of human stature: the figures are enhanced by their association with technology. *Scientific American* illustrated its cover story on the great gas meter of the Consolidated Gas Company of New York with an engraving of the massive gas drum dominating the picture while visitors are seen in much-reduced size (fig. 27). The cylinder was seventeen feet nine inches long and eighteen feet six

27 "The Great Gas Meter of the Consolidated Gas Company of New York." *Scientific American*, December 4, 1886.

Art, Technology, and the Human Image   77

28 John Ferguson Weir, *The Gun Foundry* (1866). Oil on canvas. Putnam County Historical Society, Cold Spring, New York.

inches in diameter with a drum capacity of three thousand cubic feet, but even this large size is further emphasized by the sharpened perspective of the drum receding into the distance. While the artist has glamorized the drum by exaggerating its size, the human figures in the illustration suggest that they are by no means overpowered or overwhelmed by technology: a bearded attendant in frock coat is nonchalantly seated in front of the huge metal drum, undisturbed by its looming presence.

American painter John Ferguson Weir (1841–1926) took a similarly sanguine view of massive technologies. Son of a West Point drawing instructor who later became dean of the Yale School of Fine Arts, Weir, in his painting *The Gun Foundry* (1866, fig. 28), depicted ironworkers at the West Point Iron Foundry in Cold Spring, New York, casting a Parrott rifled cannon for use in the Civil War. (One year later, Weir also painted *Forging the Shaft*, a painting, as he wrote in 1865, showing "Swarthy Smiths" forging shafts for engines "huge and pondrous—glowing hot.")[8]

As they pour molten iron in a sunken mold, the small but very muscular heroized male laborers in *The Gun Foundry*, working in a

78   Breaking Frame

29 Joseph Wright of Derby, *An Iron Forge Viewed from Without* (1773). Oil on canvas. The Hermitage Museum, St. Petersberg.

vast, cavernous space, appear undaunted by the giant ironworks, and equal in strength to the technology they are using. Weir's is a romanticized view, and his dramatic use of chiaroscuro—the contrast between bright highlights and shadows which lends theatricality to the scene—echoes some of the earliest important paintings of industrial subjects, including British artist Joseph Wright's *An Iron Forge Viewed from Without* (1773) and *The Blacksmith's Shop* (1771), where light cast by forge fires illuminates the faces of muscular male figures in the dark. Wright's late eighteenth-century painting of an iron forge (fig. 29) is an intimate view of industry where two men and a woman grouped in a small structure intently watch the labors of a bent male working on the iron. The cutaway view of the brightly lit forge interior emphasizes the small scale of the building in the context of the larger dark night landscape outside, where a moon, seen through clouds, creates shadows of the sublime. Wright's forge is a huddled outpost of industry, a place of intense focus endowed with a picturesque aura of awe and fear (the woman is stereotypically shown turned away from the process, holding on to a male companion while turning her head back to see the fire).

Painted nearly a hundred years later, Weir's *Gun Foundry* has a much greater expansiveness, reflecting the rapid industrial growth and confidence of the era. His human figures work in a great arena of technology: the enlarged space and vast height of the foundry expand to fill the entire painting. Even as the viewer looks down on the scene from a raised perspective, the workers emerge as undiminished in stature, equal to the task of pouring the huge caldrons of molten metal. Weir's paintings testify to the American vision of technological promise and prowess, revealing through size and scale the country's persistent belief in human abilities and national achievements.

MECHANIZED HUMANS AND HUMAN MACHINES

During the nineteenth century, artists' images of automatons became central metaphors for the dreams and nightmares of societies undergoing rapid technological change. In a world where new labor-saving inventions were expanding human capabilities and where a growing number of people were employed in factory systems calling for rote actions and impersonal efficiency, nineteenth-century artists confronted one of the most profound issues raised by new technologies: the possibility that people's identities and emotional lives would take on the properties of machines.

Beginning in the late 1820s, British artists, sensitive to the social implications and psychological impact of new steam inventions, particularly the steam carriage and railroad, began spoofing the fundamentally absurd implications of mechanized humans who had become overly enamored of steam-powered appendages. In Robert Seymour's etching *Locomotion* (fig. 30), an elderly gentleman happily absorbed in reading a magazine titled *New Inventions* becomes a comic, segmented figure, propelled forward with the aid of large, cylindrical, steam-powered legs and a steam boiler strapped to his back.

The prospect of mechanized humans, though, was no joke to nineteenth-century observers. Thomas Carlyle, in his essay "Signs of the Times" published in the *Edinburgh Review* in 1829, lamented that "men are grown mechanical in head and in heart, as well as in

30 Robert Seymour, *Locomotion* (1829). Hand-colored etching. The Metropolitan Museum of Art, Gift of Paul Bird, Jr., 1962 (62.696.15).

hand."⁹ Carlyle characterized his own era as "The Mechanical Age," a period in which "nothing is now done directly, or by hand; all is by rule and calculated contrivance." Recognizing that mechanization was valuable for having increased human powers, Carlyle nevertheless saw technology as a threat to human individuality. While Dunin's Expanding Man was an invention that assumed the benefits of standardized clothes-fitting, Carlyle objected to the emergence of the pervasive "new trade" of "codification," or standardization, which deemphasized individual variations in favor of the general: men, he complained, were being provided with "patent breeches" by those assuming that people did "*not* need to be measured first."

But Carlyle's deeper objection in "Signs of the Times" was to the mechanization of human consciousness. He posited a dual view of human nature comprised of the mechanical and the dynamical. The dynamical aspect of human nature included forces and energies in their primary, unmodified form, "the mysterious springs of Love, and Fear, and Wonder, of Enthusiasm, Poetry, Religion, all of which have a truly vital and *infinite* character." It was the dynamical that was being lost in an age of mechanism, and that represented the qualities of human nature most responsible for science and art—for the "great elements of human enjoyment, the attainments and possessions that exalt man's life to its present height." Mechanization, he concluded, remained a threat not only to human identity but also to human well-being, for although "mechanism, wisely contrived, has done much for man in a social and moral point of view, we cannot be persuaded that it has ever been the chief source of his worth or happiness."

Carlyle's grim specter of people taking on the psychic properties of machines at great cost to their vitality and creative emotional lives contrasted sharply with the more sanguine conceptions of machine-like humans during previous centuries, including the writings of Descartes and the eighteenth-century French philosophes, whose arguments were informed by a Newtonian vision of an orderly, mechanistic, clockwork universe. In his *Discourse on Method* and *Philosophical Letters*, Descartes argued that animals and human beings are essentially machines, with the important difference that human beings alone have a mind or soul that is not a mechanism.[10]

A more extreme view of the human as machine was presented by eighteenth-century French philosopher Julien Offray de La Mettrie, who in his essay *L'Homme machine* (Man a machine, 1748) embraced the metaphor of humanity as a well-tuned mechanism, functioning with the precision of a finely constructed watch. Extending the arguments of Descartes, who had likened the involuntary movements of the animal body to those of a machine, La Mettrie argued that "man is but an animal, or a collection of springs which wind each other up." "The human body," he wrote, "is a machine which winds its own springs. It is the living image of perpetual movement."[11]

La Mettrie's ready acceptance of a machine metaphor for the body's workings represented a tribute to human abilities in the Age of Reason—a tribute seen also in the fascination with mechanical automatons that typified eighteenth-century France. The delight in automatons extended back even earlier to ancient Egyptian moving statuettes with articulated arms, and to the automatons of ancient Chinese, ancient Greek, and medieval Arab artisans, as well as to European clockmakers of the medieval and Renaissance periods. Used for religious rituals, for demonstrating technological principles, as toys and mechanical wonders intended to delight and entertain, and as functional machines, these early automatons became central cultural images representing the mechanized human as a benign and beneficent creation.

In his treatise on mechanical designs, the seventeenth-century Italian engineer Giovanni Branca illustrated a steam boiler in the shape a human torso; a pipe extending from the figure's mouth turned the vanes of a horizontal wheel.[12] Branca's anthropomorphic boiler, a machine in the guise of a human figure, both embodies and empowers the machine, enriching rather than threatening human life.

Other historical images of humanlike machines emphasized their docile subservience to their human creators. Homer in *The Iliad* described two female automatons who aided the god Hephaestus, the craftsman for the gods. The women were "golden maidservants" who "looked like real girls and could not only speak and use their limbs but were endowed with intelligence and trained in handiwork by the immortal gods."[13] Hero of Alexandria in his treatise *Pneumatica* (c. 62 A.D.) presented automated figures in animal and human shapes which performed simple movements and actions demonstrating the applications of science, and his steam Aeolipile, in demonstrating the expansion of a gas when heated, caused figures of dancers on a turntable to move.

The idea of automatons as useful servants and amusing toys continued in the designs of medieval and Renaissance clockmakers, whose figures, deriving their movements from clock mechanisms, struck the hours, as in the metal giants atop the clock tower in the Piazza San Marco in Venice (1497). The tradition became even more elaborate in the clockwork automatons that delighted audiences in eighteenth-century France, the nearly magical human figures that could play musical instruments, draw pictures, write letters. Jacques de Vaucanson amazed the French Academy in 1738 with his automated flute player and his automated duck that could swim, eat, and "digest" its food.[14]

Rivaling Vaucanson with their complex automatons, the Swiss craftsmen Pierre Jacquet-Droz and his son Henri-Louis gained fame for a letter-writing boy scribe and an exquisitely lifelike lady musician that played the organ as her bosom heaved and her eyeballs moved (fig. 31). Another Jacquet-Droz automaton of a girl playing the harpsichord was admiringly described in a contemporary poem as "a vestal virgin with a heart of steel."[15] With their hidden mechanisms and elegant simulations of human appearance and actions, the French automatons ultimately heightened human stature through the flattery of imitation: the machines mimicked rather than demeaned the human figure, while also paying tribute to the designer's mechanical genius.

31 Pierre and Henri-Louis Jacquet-Droz, Lady Musician (1773). Musée d'Art et d'Histoire, Neuchâtel, Switzerland.

Nineteenth-century Americans shared the French delight in automatons. While Carlyle in 1829 was warning against mechanized human psyches and while nineteenth-century British caricaturists were satirizing mechanized human figures, American artists produced few if any satirical robotic images, even during a burgeoning period of steam inventions. America's artistry with human machines lay in the hands of inventors and toymakers, who delighted in creating playful versions of mechanized humans. Robert Seymour in *Locomotion* spoofed the idea of a steam-powered man, but American inventors saw the idea as a source of inspiration. Newark mechanic Zadoc P. Dederick in 1868 patented a seven-foot nine-inch steam man sporting a top hat with a body containing a three-horsepower steam engine intended to travel thirty miles an hour and pull a wheeled carriage, an idea that appeared earlier in Edward S. Ellis's dime novel

The Steam Man of the Prairies (1865) and spawned another series of dime novels starting in 1883.[16]

Beginning in the 1860s, a fertile period of American invention, American toymakers created their own beneficent versions of mechanical humans, taking out a series of patents for clockwork walking dolls. New York inventor Enoch Rice Morrison in 1862 patented a walking doll he called "Autoperipatetikos" (derived from the Greek for "self-propelled" and "walk about"), a female doll with a bisque or cloth head and a hidden, spring-driven clock mechanism that allowed it to move its arms and walk on its cut brass legs.[17]

American inventors of these automated dolls capitalized on the successful clock industry centered in Connecticut, recognizing that well-made clock mechanisms with heavy steel springs and heavy brass gears would work well as drive systems for toys.[18] The walking toy dolls with their clockwork mechanisms and die-out segmented body parts were very much a product of a society already devoted to the concept of standardized manufacture. The Connecticut clock industry and American firearms manufacture were known for their large-scale production methods based on the "uniformity system," or the American System, using division of labor and interchangeable, standardized parts.[19] In a society that sanctioned these systems, a society devoted to Yankee ingenuity—and a society that welcomed steam-powered transportation as a means to fulfill national goals of expansion and cohesion—the idea of a steam-powered or clockwork human was apt to be a source of pleasure rather than alarm.

THE MENACING MACHINE

In 1889, America's archetypal inventor, Thomas Edison, began manufacturing his own version of a mechanical human. His phonographic "talking" doll, first patented by William W. Jacques in 1888 with the patent later assigned to Edison, was made of a tin body with record cylinders inside that played a song or speech when a crank was turned.[20]

Although Edison's phonographic doll was considered another wondrous product of American ingenuity and a positive cultural emblem of an automaton, *Scientific American* in a story published in 1890 on the doll's manufacture evoked a factory scene that was unexpectedly tinged with the surreal: the many girls hired to make recordings on the cylinders virtually became automatons themselves, creating cacophony as they continuously recited their rhymed words to be repeated by the doll. The jangle produced by the girls simultaneously repeating "'Mary had a little lamb,' 'Jack and Jill,' 'Little Bo-peep'" was "beyond description." "These sounds," the magazine added, "make a veritable pandemonium."[21]

Amid a happy scene of American technological know-how, the sounds of pandemonium in the *Scientific American* story hinted at a second, darker theme in American and more often European cultural imagery, one that conjured up the unsettling disruptiveness of the industrial age and presented mechanized humans as ludicrously comic or dreaded figures. It was this view that predominated in British caricatures of mechanized humans during the nineteenth century, and even surfaced at times in the American consciousness.

One image of a menacing, troubling, artificially created human simulacrum appeared in sixteenth-century Hebraic legends of the golem. The word *golem* as it appeared in the biblical Psalms meant unformed substance, embryo, or shapeless mass, and the golem in literature was an artificially created but not mechanical creature. Sixteenth-century legends credited Elijah of Chelm as the first person to make an artificial man by invoking the divine name, yet the golem was said to be a monster menacing the world. The golem created by Rabbi Loew of Prague in 1560 was a speechless male figure shaped out of clay, used as a servant, and other golem legends were fostered by Hasidic writings after the seventeenth century.[22]

Another literary image of the mechanized human figure was poignantly presented in the guise of the beautiful automaton Olympia in E.T.A. Hoffmann's story "The Sand-Man" (1816), and in Herman Melville's story "The Bell-Tower" (1855), an errant clockwork automaton kills its inventor. But the theme received its most evocative and archetypal embodiment in Mary Shelley's gothic novel *Frankenstein*.

Shelley's novel epitomized the image of a dangerous, gigantic, artificial human with its grieving heart and bitterly murderous tendencies. Her menacing figure has been attributed to multiple sources—her own preoccupation with blighted creations based on her painful experiences with miscarriages and the death of her children; her awareness of the discovery of animal electricity in 1791 by Luigi Galvani; her knowledge of Paracelsus and his writings on the creation of a homunculus, or artificial human, cultured in a glass vessel; her visit in 1816 to Neuchâtel where she saw the automatons made by Jacquet-Droz; and her reading of the French philosopher

Condillac, whose *Treatise on the Sensations* (1754) imagined a lifelike statue capable of thinking.[23]

The "demon" or "miserable monster," as Frankenstein calls his own creation, has a "human frame" which is terrifying, evoking horror and disgust. But the damaging behavior of this monster occurs only after it is rejected and disowned, left to ravage the landscape without being befriended or tamed. Frankenstein's artificial man, mysteriously created from "lifeless matter," is not mechanical but became mythologized as the embodiment of technology gone awry. Indeed, the author's description of the monster's reappearance echoes images of industrial technology that occasionally appeared in eighteenth-century British landscape paintings. During his wanderings toward the Alpine valley of Chamonix, after his creation of the monster, Frankenstein describes the immense mountains in language suggesting the pictorial sublime and the picturesque:

> Ruined castles hanging on the precipices of piney mountains; the impetuous Arve, and cottages every here and there peeping forth from the trees, formed a scene of singular beauty. But it was augmented and rendered sublime by the mighty Alps.

The next day,

> These sublime and magnificent scenes afforded me the greatest consolation that I was capable of feeling. They elevated me from all littleness of feeling; and although they did not remove my grief, they subdued and tranquilized it.[24]

It is in the midst of feeling himself elevated from human "littleness" by the magnificent landscape that Frankenstein is diminished in scale by the sudden unwelcome appearance of the man-monster "advancing towards me with superhuman speed." The sight of his figure, whose stature "seemed to exceed that of man," causes him to feel faint before he is overcome with "rage and horror." Frankenstein's horror at his artificial man stems from revulsion at its macabre appearance, gruesome origins, and murderous impulses—and his despair at losing control of a monstrous creation that now, much like the appearance of fiery, gigantic new industries, threatens to defile the landscape.

Shelley's gothic image of horror gained potency and power from its ability to capture the nightmare of an artificially created superhuman terrorizing humanity and tormenting its creator. In her novel, it is the very element of artifice, of manufacture, of *techne* that makes the monster particularly terrifying. In an age that was busily constructing new realities from its technological prowess, here is *techne* gone wrong, technology grotesquely misapplied, resulting in human deformity and defacement.

Shelley's monster was not only an archetype of technological threat but also posed troubling questions of human definition in a technological context. The monster figure with its "human frame" made from charnel house bones, dissecting-room and slaughterhouse flesh becomes a morbid travesty of human identity, a "filthy mass" as Frankenstein calls him, an alien outcast in the human landscape. Frankenstein himself ultimately becomes dehumanized by his obsessive drive to destroy the monster, and by the end of the novel he himself, transformed by rage and passion, has become something of a mechanical man: "I pursued my path towards the destruction of the daemon more as a task enjoined by heaven, as the mechanical impulse of some power of which I was unconscious, than as the ardent desire of my soul."[25]

The challenge to human identity posed by Shelley's technological simulacrum raised issues that haunted nineteenth-century thinking even as the focus sharpened to more particularized fears of mechanized humans. It was the idea of people themselves taking on the properties of machines—by wearing mechanical prostheses, by relying on technological aids, by becoming robotlike in their jerky movements and emotionless responses—that became, to some writers and artists, a source of amusement, scoffing, and alarm.

In scientific circles, a hint of this undercurrent of anxiety appeared in "The Artificial Man," an essay published in an early volume of *Scientific American* (1859) and reprinted from a British journal. The author recounts his visit to a medical library where he catches sight of a volume titled *Biggs on Artificial Limbs*, which describes the work of a London manufacturer of prosthetic devices.[26] Visiting the manufacturer's workshop, the author is shown a range of legs, eyes, noses, and teeth, which, in this society sensitive to social class, come in various versions ranging from no-nonsense, workmanlike limbs to elegant models intended for the aristocracy. An accompanying illustration jauntily depicts one of the firm's workers with an artificial leg slung over his shoulder and a severed hand on a table nearby.

The author's visit to the limb factory leads him to troubling hallucinations as he walks down the street wondering if the people he encounters are genuine or made of artificial parts: "As I walked homeward, my head full of the subject I had been dwelling upon, the 'artificial man' seemed to meet me in detail everywhere." Contemplating this hallucinatory vision of artificial teeth and wigs everywhere, he asks, "What member is there in this artful age that we can depend on as genuine? What secret bodily defect that we particularly desire to keep to ourselves that that wicked *Times* does not show up in its advertising sheet and tell us how to tinker?"

The description becomes comic-macabre as the writer begins to imagine the dismemberment of the "artificial man": "If the individual can thus craftily be built up, imagine, good reader, the nightly dissolution. Picture your valet taking off both your legs . . . carefully placing away your arm, disengaging your wig, easing you of your glass eye, washing and putting by your masticators, and, finally, helping the bare vital principle into bed, there to lie up in ordinary, like a dismantled hulk, for the rest of the night!" The article ends with a telling lament and prophecy as it conjures up the image of Shelley's monster: "In these latter days we are, indeed . . . fearfully and wonderfully made; and, like the author of 'Frankenstein,' we may tremble at our creations."

It was this funny and disturbing vision of mechanical man, whose identity was being systematically built up and then unglamorously taken apart into a "dismantled hulk," that was made vividly and sometimes hilariously concrete in the work of nineteenth-century British caricaturists.

A PROSTHETIC GOD

The nineteenth century's infatuation with steam technologies and labor-saving devices became a source of comedy to caricaturists who perceived that through the use of these inventions people were, in effect, becoming machinelike themselves. Britain's Robert Seymour, William Heath, and George Cruikshank, along with Honoré Daumier, Cham (Amédée de Noé), and Albert Robida in France, were among those artists whose engravings, etchings, and lithographs—for all their seemingly lighthearted appearance—recognized the tragicomic implications of people assuming the robotic sensibilities of human automatons. As Henri Bergson observed in his study *Laughter* (1900), there is a strong element of comedy when the artificial is substituted for the natural, when the mechanical is layered on the living: "The attitudes, gestures and movements of the human body are laughable in exact proportion as that body reminds us of a mere machine."[27]

The eighteenth-century French artisans who produced automatons elicited admiration for having created machines that imitated the appearance of humans, and the mechanisms were generally hidden from view. There was nothing absurd about these mechanized human imitations: they drew smiles of praise rather than skeptical smiles of disdain. But artists and craftsmen also recognized the implicit absurdity of human beings who, often unwittingly, appeared as machines. In this second version of the mechanized human, machine parts were usually exposed and even emphasized, and the effect was one of comic deflation: the sight of human beings with machine parts for heads, arms, and legs could easily be viewed as freakish and bizarre. One of the earliest etchings of a human made of machine parts was by seventeenth-century Florentine artist Giovanni Battista Bracelli, who named his series of forty-five etchings *Bizzarie di varie figure* (1624). Bracelli's stick figures were acrobats, dancers, knife grinders—metallic men made of machine parts.

Later humorists recognized that mechanized humans were funny (or morbid) to the extent that the artifice was obvious and the machine parts were grafted on disjointedly. Frankenstein's monster, the artificially created human, was monstrous because verisimilitude failed: his mummy face, his waxy yellow skin, testified to his necrological origins—from parts of dead bodies joined together by Victor Frankenstein into an uneasy and unholy alliance.

In a less macabre vein, nineteenth-century caricaturists enjoyed ridiculing the public's love affair with mechanical devices, an infatuation that was turning people into shameless automatons who wore their mechanical hearts on their sleeves (or their legs, as the case might be). Some of the liveliest satires came from Britain's Robert Seymour. Before turning to caricature, Seymour ("Shortshanks") had received training as an oil painter, and his early painting of Don Quixote presaged his penchant for satirizing dotty human foibles and fantasies. His caricatures of technology published by Thomas McLean began appearing in 1829, the period that coincided with early steam carriage experiments in London.

Seymour's skilled satires of technology presented people who had taken on the look of machines by being thoroughly and often literally wrapped up in new mechanical devices, oblivious to their silly appearance. Beneath the trappings of farce, however, Seymour's art reflected the recurrent nineteenth-century fear that people were becoming so reliant on machines that they were losing their human capabilities. During the period when steam engines were rapidly being adapted to drive carriages, ships, and locomotives, many welcomed the improved means of transport. Yet for Seymour, the new steam-driven inventions undermined human self-sufficiency. Seymour's first version of *Locomotion* satirized not only the elderly gentleman in his steam-powered mechanical legs but also a man flying a steam boiler with wings and women riding in a teakettle-shape carriage powered with an infusion of "gun-powder tea." (The image of a man in a flying steam boiler came close to reality in 1843, when British inventor William Henson patented an "Aerial Steam Carriage," a propeller-driven, fixed-wing monoplane powered by a thirty-horsepower steam engine set in a fuselage. In actuality, the machine failed in flight.)

Seymour's images of technologies satirized the human impulse to augment one's own body and also prophesied an overdependence on technology, a prospect that later worried Freud, who in *Civilization and Its Discontents* warned about mechanical prostheses:

> Long ago [man] formed an ideal conception of omnipotence and omniscience which he embodied in his gods.... To-day he has come very close to the attainment of this ideal, he has almost become a god himself.... Man has, as it were, become a kind of prosthetic God. When he puts on all his auxiliary organs he is truly magnificent; but these organs have not grown on to him and they still give him much trouble at times.[28]

Seymour's two versions of *Locomotion* reflected his awareness that mechanical prostheses were indeed troublesome, and his engravings became compendiums of mechanical hazards with their attendant anxieties.

The artist's second version of *Locomotion* (fig. 32) expanded his vision of balking machines. Another gentleman encased in steam-powered legs has become stalled and asks, "Curse the fire—is it out?" as a kneeling attendant works anxiously at bellows to fan the flames. In the distance, travelers are seen being ejected from a steam carriage bearing the ironic legend "Safety Coach—Warranted not to Explode." Overhead is additional evidence of an anxiety-provoking technology: a man astride steam boilers flying in his fantasy machine is stranded up in the air as he calls out, "Help, help, the joint won't turn and I can't get down again!" For all its comic appearance, Seymour's etching highlights somber human fears of being paralyzed,

Art, Technology, and the Human Image 91

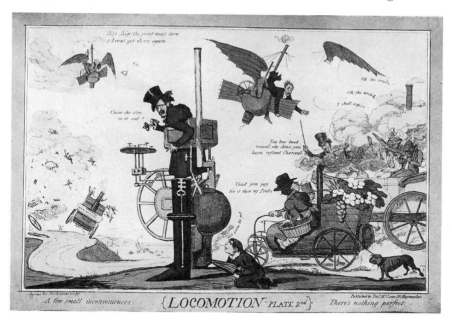

32 Robert Seymour, "A Few Small Inconveniences. There's Nothing Perfect," plate 2 of *Locomotion* (c. 1830). Etching. Reproduced by courtesy of the Trustees of the British Museum.

stranded, cut off from earth, shattered by explosions. Belying his serious subject, the etching is captioned, "A few small inconveniences. There's nothing perfect."

Seymour's satires of mechanical, steam-powered legs suggest also a curious type of technological metonymy with individual parts used to suggest the whole. The mechanical human body parts are absurd to the extent that they become autonomous, taking on a life of their own. These mechanical appendages can speed up or (as in Seymour's caricatures) retard human movements, but it is their recalcitrance, their independence from human control, that can change them from life-enhancing to life-destroying devices.

A nineteenth-century London music hall song, "The Steam Arm" by composer Henry Miller, capitalized on the comic image of a runaway appendage, again suggesting the familiar theme of technology out of control. The song tells of a soldier who, after losing his arm at Waterloo, has an arm glued on which is worked by steam and has a murderous punch that fells the town policemen and mayor: "For the soldier's arm had been so drill'd, / That once in action, it couldn't be still'd." As the power of the man's arm grows, so do the violence and acts of destruction. Ultimately, he "wanders about just like a sprite; / For he can't keep still, either day or night, / And his arm keeps moving with two horse might."

If the steam arm lands the soldier in a purgatory of restless wandering, the mechanical legs and flying machines in Seymour's *Locomotion* prints thwart their wearers not by rushing them on but by causing a frustrating immobility. These balky devices, as Seymour writes, cause "inconvenience" and perhaps some loss of human dignity. But, as seen in the exploding engine, they also threaten total human disintegration: the adding on of mechanized appendages has, as its nightmare antithesis, brutal dismemberment. The explosion of severed body parts, a visual metonymy for technology seen even more clearly in Seymour's print *Going It by Steam* (fig. 14), becomes an act of deconstruction: the grotesque, unexpected disassembling of mechanical man.

Seymour's comic view of vulnerable travelers victimized by mechanical inventions highlighted caricaturists' amusement with labor-saving devices. The nineteenth century also had its forerunners of twentieth-century American cartoonist Rube Goldberg, who delighted in inventing comic contraptions designed to ease the burdens of daily living. In their own visions of fantasy machines, British caricaturists revealed a lurking concern that mechanization was actually undermining human capabilities, producing a dangerous laziness. *Living Made Easy*, for example, a series of etchings by Seymour published by Thomas McLean in 1830, pictures an "Apparatus to Undress and Cover Up While Sleeping" that pulls off a man's trousers and puts on his nightshirt. In a state of luxuriant ease, a character in another engraving is seen reclining on a red velvet chaise while being fed nuts from a bowl and wine from a bottle, held by a machine labeled "Body-Fanner, Nut-Cracker and Wine Helper, for the Heats of Summer" (fig. 33).

The most creative and pictorially complex assaults on people with their mechanical labor-saving devices appeared in a series of prints titled *The March of Intellect* by Seymour and by British caricaturist William Heath. Both artists' antic images captured the foibles of people changing their life styles and identities with the aid of new machines. "The March of Intellect," a popular phrase in British culture after 1825, referred in part to the wider availability of knowledge about technology to laboring classes, a version of progress that became a source of humor for caricaturists. These satiric engravings

Art, Technology, and the Human Image 93

33 Robert Seymour, "Body-Fanner, Nut-Cracker and Wine Helper, for the Heats of Summer," *Living Made Easy* (1830). Hand-colored etching. Print Collection, Miriam and Ira D. Wallach Division of Art, Prints, and Photographs, The New York Public Library, Astor, Lenox and Tilden Foundations.

and etchings, published in several versions in 1829, contained multiple cartoon vignettes in a single print, and the format itself became, in effect, an aesthetic strategy for taming technology, for putting a boundary or framework around the forces of rapid technological change. Confined and ordered within the borders of a print, unruly technological developments with their attendant human fears and anxieties were mirrored and, in a sense, made manageable.

Heath's best engravings on the subject of technology were his two prints titled *The March of Intellect* (1829). In one version, Heath encased a host of contemporary steam inventions within the framework of a five-story house; a cutaway view of the interior is captioned "Acme of Human Invention. Grand Servant Superseding Apparatus for Doing Every Kind of Household Work." The print satirizes the fact that by 1829 steam engines were being used not only for transport but also for inventions ranging from the practical to the preposterous. Heath's engraving, which spoofs the lives of the idle rich, reflects the artist's skeptical position toward these mechanical labor-saving devices, implying that they induce laziness and dependence. His fantasy machines are both absurd and prophetic; they include drawings of a self-discharging coal shuttle, a dressing machine, and an automated washing and ironing machine with a series of irons strung on a moving pulley.

NEW HUMAN CONTOURS

Robert Seymour's *The March of Intellect*, subtitled *The Mechanical* (1829, fig. 34), was one of his series of six etchings lampooning the social fashions of his time. The print reflected a complex range of feelings about the way technology was restructuring the very contours of human identity—a belief that was, paradoxically, both a hope and fear. The human figures in Seymour's print are all engaged in a mechanically aided reshaping of their bodies. In a mock identity-changing salon, one man sits in a tub filled with "grand softening fluid" while another patron of short stature asks the proprietor, "Can you make me tall and genteel like my son Bobby here?" Another vignette shows a man in the grips of a torturing racklike device labeled "stretching machine." In one of the more unnerving comic views, a salon patron is seen being subjected to a process advertised as "Amputation by Steam. No pain!!" Seated with his arm in the hole of a wooden contrivance, he cries in alarm, "Hello—You've cut off my arm instead of my leg!"

Seymour's comic references to amputation and dismemberment reflected not only the literal dangers posed by new machines but also deeper fears of being entrapped, tormented, desexed by new technologies—anxieties that would later emerge as important themes in Dada art and among German Weimar artists during the 1920s and 1930s. The stretching of the human figure in Seymour's etching also, of course, had optimistic overtones: it reflected the popular conviction that mechanization would heighten human stature and expand human progress—as was later seen in the stretched human figure of Dunin's Expanding Man at the Great Exhibition of 1851.

In the guise of lighthearted cartoons, Seymour's prints shrewdly captured the somber implications of new technologies. Seymour's characters have a peculiar naïveté as they again and again find themselves trying out new machines only to have the worst happen. *Shaving by Steam* (1828, fig. 35) shows men who have come to participate in the mundane act of having their beards shaved but discover that the experience touches on their most profound fears. Barbershop customers sit with their heads in a series of wedges, awaiting their turn at a "shavograph," an elaborate contraption consisting of

34 Robert Seymour, *The March of Intellect, The Mechanical* (1829). Hand-colored etching (detail). Ironbridge Gorge Museum Trust.

35 Robert Seymour, *Shaving by Steam* (1828). Engraving. Reproduced by courtesy of the Trustees of the British Museum.

brushes, lather makers, and razors all moved by cogwheels and attended by an engineer with a directing rod. A fashionably dressed woman stands at a counter. As several customers seated on the left await their turn, they are unmindful that a fellow customer gesticulates wildly as he discovers that a steam razor has just sliced off his nose. He holds out his hands before a stream of splashing blood as the anxious next man in line insists, "Stop! Stop!"

Seymour's engraving again evokes fears of unstoppable technology, a technology that may inadvertently have a drastic impact on human identity. The engraving also suggests, in a sense, an even graver fear of castration by machine. The man, as the caption to Seymour's engraving explains, with mock moralism, ignored the injunction to sit "firm and steady" with his nose in the machine's wedgelike hole. Making the mistake of being distracted by the woman at the counter the victim must pay "the penalty of his own imprudence." Seymour's "shavograph" is finally, for all its other meanings, the image of remorseless technology. Machines may be designed to manipulate, channel, and tame Mother Nature, but they also wreak havoc, reducing their users to a state of helpless impotence.

The comically threatening implications of Seymour's shavograph contrast significantly with an earlier spoof of a shaving machine drawn at the end of the eighteenth century. *The Shaving Machine* (1794, fig. 36), a satiric engraving by an anonymous British artist, purports to be a detailed illustration of an apparatus manufactured by "D. Merry and Sons" of Birmingham. The plate shows six customers seated in a shaving and wig-dressing room, with two men being soaped and brushed by a machine holding a line of razors and brushes that can shave up to twenty customers at a time. The machine is driven by a large crank-operated pinion wheel turned by an attendant at the side, while a boy in the foreground is dressing a wig and another man with a hair-powdering gun is shooting powder at a row of wigs. Refuting the charge that "the ingenious mechanics of Birmingham" lack "invention and originality," the print's written text solemnly answers those who fear for their safety, reassuring skeptics that the machine will "do its work in the most safe, smooth, and efficacious manner, with three scrapes or movements" (though the master of the shop tells a man with a large nose to keep it out of the way).

Playfully tweaking the reader's nose in this age of ingenuity, the text also claims that the ambitious inventor of the shaving machine expects to produce another machine for combing, curling, "frizling," and powdering hair which will powder several wigs at a time. Again, spoofing contemporary anxieties about technological mishaps, the

36 *The Shaving Machine sold by D. Merry and Sons* (Birmingham) (1794). Engraving and letterpress. Trustees of the Science Museum (London).

text confides that while the manufacturer was perfecting his new hair-curling machine, the boy who was heating the stove connected to the curling irons accidentally set the machinery on fire.

The *Shaving Machine* engraving of 1794, with its straight-faced mimicry of contemporary technological drawings (the parts of the machine are all carefully labeled), captures the dialectic of the age: the fascination with technological ingenuity and the underlying fear of dangerous accidents. The invention also, in its own way, represents a challenge to human identity. For all of its promised efficiency, it is, after all, a very impersonal mechanism that treats all customers alike and even poses a threat to their (very vulnerable) noses. *The Shaving Machine* also raises the issue of identity by suggesting yet another feature of the machine: the manufacturer tells of his idea to add a "parallellogram" and a linseed oil–turpentine mixture to his shavograph, enabling the hair dresser to make a perfect likeness or copy of the customer's face superior to the black cut-paper profiles "done by mirrors," so that a "Gentleman might be shaved, dressed, and have his portrait in the bargain, for the easy charge of sixpence."

The portrait-making mechanism is a forecast of the fascination with duplication and replication in the fine arts which would increasingly preoccupy the next century. It also represents a devaluation of human personality and uniqueness: it is an impersonal technique that eliminates not only the subtlety, and perhaps insight, of the artist's hand but also any possible sensitivity to the subject's interior life.

This theme of depersonalization and devaluation of the individual provoked by new technologies became an important theme in nineteenth-century caricatures. To the eyes of the French artists Honoré Daumier and Cham, there was something both humorous and humbling in the ways that standardized machines were maladapted to variations in individual bodies, leading to situations which were both comic and dangerous. One of the favorite targets for satire was the turnstile. With its spoked form resembling a machine cog, the turnstile was an apt emblem for technology, symbolizing its tendency to regulate human activity, to make it machinelike. Daumier and Cham lampooned the tribulations of luckless people ensnared by the turnstile during their visits to the Exposition Universelle, a world's fair held in Paris in 1855 displaying recent developments in science and technology. In Cham's rueful view, the turnstile with its standardized space between spokes became technology as trap, a device designed to impede human passage rather than make it easier. In

37 Cham (Amédée de Noé), Man in Turnstile, *Promenades à l'Exposition*. *Le Charivari*, June 25, 1855. Collection of the author.

one print, a corpulent gentleman is caught in the mechanism and anxiously tells his friend, "These machines aren't designed for people of every size. You shouldn't have forced me in before making sure you could get me out" (fig. 37).[29]

Technology was equally hard on nineteenth-century women, for the new machines were not designed to accommodate their wide-skirted fashions or varying body types. In a double-barreled satire aimed at contemporary women's fashions as well as technology, caricaturists made fun of the petticoats beneath the voluminous skirts. Daumier's "Effects of the Turnstile on Crinoline Petticoats" mercilessly derides a group of haughty female visitors to the Paris Exposition who stand with their large noses pointed into the air, though their hauteur cannot hide the fact that their fashionable crinolines have been rudely deformed by the implacable turnstiles.

THE ENCIRCLING MACHINE

The French exposition visitors comically ensnared by the spokes of the turnstile represent a recurrent pictorial image in nineteenth-century art: the integration of a human figure with a circular emblem of the machine. The human figure is juxtaposed next to, or merged with, a large wheel, rotor, cog, or cylinder—a circular mechanical shape that becomes a pictorial metonymy for industry or technology. These pictorial images suggest a troubling merger of identities between man and machine, with the human figure and psyche becoming modified if not unalterably transformed by the demands of machine production.

John Ruskin, in "The Nature of Gothic," feared the depersonalizing impact of industrial labor, arguing that "men were not intended to work with the accuracy of tools, to be precise and perfect in all their actions. If you have that precision out of them, and make their fingers measure degrees like cog-wheels, and their arms strike curves like compasses, you must unhumanize them."[30] This troubling view of human machines has its literary counterpart in Melville's grim story "The Tartarus of Maids" (1855), in which the narrator, visiting an American paper mill, has a hellish vision. The young girls working in the mill "served mutely and cringingly as the slave serves the Sultan," and their identities have become diabolically debased and fused with the machine: the girls "did not so much seem accessory wheels to the general machinery as mere cogs to the wheels."[31]

Melville's image of the human figure reduced to a mere cog in the wheel contrasts sharply with a more hopeful conception of human potential presented centuries earlier. To the Renaissance artist on the eve of the scientific revolution, human capabilities seemed largely unbounded. In Leonardo da Vinci's *Proportions of the Human Body* (1485–1490), a nude male figure stands with arms and legs outstretched, his limbs touching the circumference of a circle and square which surround him. Leonardo's man not only embodies the artist's notion of ideal human proportions but also encapsulates Renaissance views of human potential: through his intelligence, invention, and reason, man can heighten his stature by reaching out to understand, control, and master the forces of nature. A quintessential Renaissance figure, Leonardo was both artist and engineer, his probing intelligence directing him to analyze both the natural world and mechanics. Seen in Leonardo's notebooks are drawings of human anatomy and plant life as well as inventive war machines. Fascinated with the potential for human flight, he observed the flight of birds to help him design ornithopters—human-powered flying machines emblematic of his quest to transcend human, earth-bound limitations.

The juxtaposition of man and circle in Leonardo's study is a happy integration, an emblem of human aspirations in the intimately connected realms of art, nature, science, and technology. But nearly four centuries later, nineteenth-century artists commenting on the impact of mechanization and technology had more ambivalent views of the integration of human and machine. While Renaissance artists took great pride in human intelligence and abilities, these nineteenth-century artists' images of the human figure were often more circumspect. Rather than extending their reach to test human limits, these figures were put on a circular rack, often uncomfortably circumscribed or crowded into the encircling confines of their own technological creations.

To nineteenth-century artists with political agendas, the image of the encircling machine would take on an added dimension of torture and torment. Jean Graves, anarchist editor of the radical French newspaper *Les Temps Nouveaux*, frequently published the art of Camille Pissarro, Franz Kupka, and others who protested against the dehumanizing effects of industrialization. Kupka's print "The Cog," published in *Les Temps Nouveaux* in 1905, presents a haunting view of human victimization by machine as a laborer, his arms and legs chained within the confines of a cog, is symbolically crucified by his enchainment to a factory system of production.

While politicized illustrators used the encircled human figure to suggest the tortured and dehumanizing impact of technology, the visual image of a factory worker completely framed by an encircling cog, wheel, or cylinder also became a more neutral pictorial convention in prints and photographs, a convenient visual device for showing workers in industrial settings, seen in "The Lathes" published in *Harper's Monthly* (fig. 38). Rather than representing an alienation of worker from machine, such images suggest a comfortable sense of cohesion and integration.

Similar emblems would continue to appear in twentieth-century images of technology, where they would retain their pictorial ambiguity, suggesting a pessimistic view of human identity deformed by

38 "The Lathes," Novelty Iron Works (detail). *Harper's Monthly*, May 1881. Photograph: Smithsonian Institution, Washington, D.C.

technology yet also the more optimistic Renaissance view of expanded human potential. In his many photographs of industry taken during the 1920s and 1930s, American photographer Lewis Hine drew on existing pictorial idioms while intensifying the merger of man and machine. His dramatic photographs of workers and rotors emerged as powerful emblems of the profoundly intimate human involvement with technology and mechanization. One workman in a Hine photograph of 1921 appears within a small circle at the center of a huge transformer rotor that fills the entire picture space. The massive machine, in Hine's view, appears glamorized as a modern-day mandala—becoming a mechanical rose window with the workman comfortably at the core of the new religious icon.

In his photograph *Powerhouse Mechanic* (1920, fig. 39), Hine presents a steamfitter tightening the bolts on a steam pump, his bent, taut, muscular figure posed within the circular outline of the machine and his head abutting the edge of the circle. The photograph creates visual tension by provoking dual readings: on the one hand, there is aesthetic satisfaction, a feeling of coherence produced by the image of unification, the good fit of man and machine. But the hu-

39 Lewis Hine, *Powerhouse Mechanic* (1920). Photograph. Courtesy International Museum of Photography at George Eastman House.

man figure also appears cramped, confined within the circumference of the wheel, making him a victim, perhaps of the industrial system. His muscularity gives him heroism and power, but his symbiotic fusion with machine bespeaks bondage, a loss of self. Hine's photographs ultimately emerge as forceful embodiments of the heroic, idealized vision of labor found in photography and art of the 1930s, yet they also reflect the legacy of nineteenth-century art with its dialectical view of human beings both ennobled and enslaved by modern technologies, their identity inextricably and often ambivalently fused with that of the machine.

Chapter 3

Technology and the Design Debate

Traveling through the English countryside past smoky industrial sites in the Midlands, the rider sitting atop a coach in the opening passages of George Eliot's novel *Felix Holt, the Radical* (1866) nostalgically remembers the once tranquil pleasures of travel by coach, that "slow, old-fashioned way of getting from one end of our country to the other"—a ride on which passengers could enjoy a view of "ruby-berried nightshade" and "oak-sheltered parks." But in the age of speeding railroads and macabre death by train, the coachman now, "as in a perpetual vision," sees a "ruined country strewn with shattered limbs."[1]

Eliot's evocative contrast between an idealized, coherent, preindustrial world and the railroad era's world of fragmentation and "shattered limbs" illuminated the tensions accompanying the nineteenth century's revolution in transportation, the sense of a sharp break between tranquillity and trauma. This schism between the old order and a new, fragmented industrial era was also seen in another realm of technology: as the editors of Britain's *Art-Union* wrote in 1848, for more than a century, "a great and silent revolution has been taking place," a revolution in the "production and reproduction of works of Art."[2]

As seen in the production of pottery at Josiah Wedgwood's North Staffordshire works and in Matthew Boulton's mechanized production of silverplate and other metalware at his Soho works, eighteenth-century manufacturers had already busily engaged in mass-producing the decorative arts. The nineteenth century brought an added boon with the "imitative arts"—the die-stamped, cast, machine-punched, embossed, and plated metal reproductions and imitations of richly ornamented decorative wares. Using new materials, technologies, and machine tools, manufacturers mimicked the look of sumptuously decorated, costly, original works of art so revered in a century devoted to the proud display of home decorations.

By midcentury, American and British manufacturers were mass-producing cast-iron architectural ornaments, sculpture, and furniture for the garden and home. Through new developments in electrochemistry, they were also successfully marketing decorative electroplate, an improved technique for plating decorative wares ranging from tea service to ornamented soup tureens. Art reproductions benefited from the new process of electrotyping, developed almost simultaneously by scientists in Russia and England in 1837, which produced copies of sculpted artworks, often museum masterpieces, through the electrodeposition of metal on molds.

The flood of imitations and reproductions produced spirited design debates as nineteenth-century critics praised and anguished over the imitative arts. Their feelings of both rapture and repulsion centered on several controversies: the moral and aesthetic impact of producing facsimiles and imitative copies of original works of art; the social impact of making the imitative arts available to a middle-class market; and the question of finding an appropriate design aesthetic for an industrial society.

The deluge of new factory-made decorative arts brought a troubled sense of split or separation to English art critic John Ruskin, who in his lectures and writings including *The Seven Lamps of Architecture* (1849) saw a profound division between the inferior, machine-made imitations, with their inherent quality of falsity and deceitfulness, and the morally and aesthetically superior handmade originals. Extending his objections even further, Ruskin saw a more fundamental gulf between art and industry itself. When asked to lecture to the Sheffield Society of Artists, he bluntly refused, arguing that the production of art in any industrial setting was impossible. "You can't have art where you have smoke," he declared, adding, "you may have it in hell, perhaps, for the Devil is too clever not to consume his own smoke if he wants to. But you will never have it in Sheffield."[3]

Ruskin's bleak assessment of the possibilities for art in the British manufacturing city emphasized his unalterable conviction that art and the machine were mutually exclusive. The very nature of an industrial environment also made good manufacturing design a virtual impossibility. Artisans who in earlier historical periods had been capable of wonderful achievements were now transformed into debased creatures incapable of elegant design. England's workmen, he insisted in his lecture "Modern Manufacture and Design," presented in 1859, were "totally destitute of designing power," a result of their living in the midst of factory-blighted nature and "unbeautiful things."[4]

While Ruskin saw only a rupture between art and manufacture, there was another important current in nineteenth-century design discourse that saw a possibility of reconciliation. Artists, it was hoped, would help assimilate the new manufactured wares by lending them a look of taste and elegance. In 1848, the British journal *Art-Union* argued for closer ties between manufacturers and artists: rather than seeing a schism, the journal urged artists to lend their talents to industrial design and form "more intimate relations" with manufacturing.[5] The British painter Joseph Wright of Derby had already shown, a century earlier, his intense fascination with the new science and industry and his willingness to lend his talents to the world of manufacturing design. His painting *An Iron Forge Viewed from Without* (fig. 29), with its dramatically illuminated cutaway view of an iron forge in a rugged barn at night, and his *Experiment on a Bird in the Air Pump* (1768) stripped away any barrier separating art from science and industry: his paintings peer into the shadowy interior, forming an intimate link to the world of experiment and technological change. The linking of art and design took place in Wright's own life as well: his art and his scientific interests intrigued Josiah Wedgwood, and the artist supplied the manufacturer with designs and models for ceramic wares.

In its articles on manufacture published in 1848, England's *Art-Union*, soon renamed the *Art Journal*, further explored the need for artists to help forge an aesthetic for the new manufactures. The journal's views on the relations between artists and manufacturers, and its attitudes toward factory-made art reproductions, highlighted two of the important issues debated during the century: the artist's role in helping to assimilate these new technologies, and the legitimacy of the machine-made "imitative arts."

Though Ruskin was preoccupied with the degrading impact of industrialization and saw little hope for artists who worked in the world of industry, the *Art-Union* in 1848 praised the idea. Reflecting the status issues of the age, the journal reassured artists that they would not be lowering themselves if they worked for industry: "There is, then, nothing derogatory to the highest Art in lending its aid to decorate objects of utility . . . the sculptor does not lower his position when he supplies a model for the moulder in iron . . . or any other substance in which cast may be taken." In fact, "the multiplication of the copies of superior Art," the journal added, "so far from degrading the artist, elevates him" from that degraded position he would have fallen into had he not adapted "the altered circumstances of the age."[6]

Nineteenth-century writers on design often used this rhetoric of elevation and degradation: the new "art manufactures" were seen as representing the heights and depths of human capabilities. Echoing one of the central assumptions of the century, the journal suggested that rather than contributing to art's degradation, the artist working for industry would help by "contributing to the moral and social elevation of his country." "Let us not be misunderstood," the journal added, "we do not wish artists to become the servants of manufacturers; we do wish them to become their friends and allies; their partners in educating the people; in improving the tastes, and consequently the morals, of the community."[7]

These twin themes—Ruskin's sense of rupture and the *Art-Union*'s hopes for reconciliation—emerged as dominant responses to the new technologies throughout the century. Those who saw signs of rupture felt that the new mass-produced cast iron and plated metals had abruptly broken with historical traditions of handicraft. They bemoaned the degraded aesthetic quality of machine-stamped ornament, the moral indefensibility of putting expensive-looking plate over base metal, and the mindless clutter of pseudohistorical ornament on manufactured articles.

Ruskin's complaints about design were also based on his disgruntled view of the mass production methods used to manufacture ornamental metalwares. Josiah Wedgwood, at his North Staffordshire Etruria ceramic works opened in 1768, had already engaged in mass-producing pottery using a division of labor, skilled craftsmen, and mechanized methods for grinding flints and mixing clay, along with transfer printing instead of hand painting, but it was the cities of Sheffield and Birmingham that drew critics' ire, often becoming emblems for vulgar designs and shoddy goods produced by mass production. By the late eighteenth century, the cities had become known for their manufacture of silverplate and factory-made imitations of handcrafted originals—manufacture using a division of labor system and stamped-out, standardized parts.[8]

Though steam power was little used before 1800, Matthew Boulton's Soho foundry near Birmingham had adapted crafts techniques to mass production. Using water and animal power, Boulton's foundry, completed in 1766, made extensive use of mechanization for duplicating decorative objects: iron and steel dies were used in stamps and presses, and in the 1770s fly presses helped mass-produce stamped decorative ornament on silverplated wares.[9] As Eric Robinson has demonstrated, the fly press became an essential tool in the mass production of decorative wares in the silverplate industries in both Sheffield and Birmingham. A tool giving a "hard, accurate, perpendicular blow," it was used not only for cutting and piercing but also for molding, stamping out candlestick parts, and producing ornamental details including the flowers and leaves on household decorative objects.[10]

During the next century, manufacturers in both England and the United States expanded the range of the decorative arts with their ornamental iron castings, electrotypes, and electroplated wares. The introduction of these new imitative arts further separated a number of influential British art critics from nineteenth-century manufacturers. While electroplate manufacturers (including George Elkington in England and Reed & Barton in America) and American cast-iron manufacturers tended to minimize the distinction between factory-made imitations and handcrafted originals, British critics, notably Augustus Welby Pugin, John Ruskin, and William Morris, insisted on maintaining the fundamental division between quality handcraft traditions and the debased imitative arts.

The manufacture of art reproductions, metal plating, and machine-stamped ornament were actually not new, as Cyril Stanley Smith has reminded us. While in the fine arts, the artist generally strives to create a unique object, the idea of duplication and reproduction has always been fundamental to the decorative arts, which are "intrinsically repetitive," for they have "a quantitative requirement, namely the imperative of covering large areas or making large numbers of individual objects." Historically, as Smith noted, these imperatives "were strong incentives to mechanization." There was already ample historical precedent for mass-producing art reproductions including bronze copies of sculpted originals. Metal plating, too, had a long tradition dating from ancient practice. And while nineteenth-century critics objected to machine-stamped ornament, dies also had long been used to stamp patterns on metal pieces such as coins and decorative buttons (Leonardo engaged in mechanized metal stamping with his wedge and screw press for stamping coins).[11]

Still, there were persistent fears about the rising flood of machine-made copies imitating original works of art. French scientist R.A.F. de Réaumur in 1722, describing the process of making cast iron malleable in order to give cheap cast-iron work a finish that imitated the look of more expensive wrought iron used for artistic purposes, voiced his misgivings about the possibility that "the number of objects we call 'beautiful' and which are simply beautiful," might be "increased beyond a certain limit." He added, "we should look with less pleasure and interest at the paintings of the great masters if daubers discovered how to paint similar ones."[12]

During the next century, Britain's *Art-Union* addressed the conundrum of whether copies made from molds threatened the integrity of handcrafted works made by skilled artisans. In its article "Carving by Machinery" published in 1848, the journal suggested that factory-made copies would not devalue the original work of art and quoted from the British craftsman and inventor C. J. Jordan, who made a clear distinction between an imitative copy and the original. Jordan, whose wood-carving machines patented during the 1840s made copies of ornamental wares by following the contours of molds and who exhibited his elaborately machine-carved Gothic revival church screen at the Great Exhibition of 1851, wrote, "I believe that machinery will do for the sculptor and carver what engraving has done for the painter," while denying that the machines would result in unemployment for artists and artisans. "Machinery cannot do the work of the mind," he insisted, "although it can assist very materially in copying it." Machines, furthermore, could not reproduce the "smoothness of surface and delicacy of finish, requisite in good works."[13]

Offering further reassurance to artists and craftsmen, Jordan claimed that his machines were not intended to emulate handcrafted work, nor would it even be "desirable to attempt it," since such machines would prove "more expensive than hand labour" and would contravene their basic purpose—to "produce the work quickly and cheaply."[14]

But for all these reassurances, nineteenth-century manufacturers became increasingly intent on blurring the distinction between the mass-produced reproduction and the original, even as critics throughout the century continued to debate the degree to which these copies were undermining the privileged role of the artist and threatening the integrity of the original work of art. Many critics had mixed reactions about incorporating these factory-made reproductions and imitations into the respected world of the decorative arts. Manufactured imitations, they insisted, could never replace the original.

For John Ruskin, the inherent dishonesty of the imitative arts, what he termed "Operative Deceit," was a "sufficient reason to determine absolute and unconditional rejection" of all cast architectural ornament and machine-made ornament. Only handmade ornament had integrity—all else he denounced as false and "utterly base."[15] The ease of creating imitations, he argued in *The Seven Lamps of Architecture*, sullied the sanctity of traditional methods of handcrafted creations, and the result was aesthetic and moral superficiality:

> one thing we have in our power—the doing without machine ornament and cast-iron work. All the stamped metals, and artificial stones, and imitation woods and bronzes, over the invention of which we hear daily exultation—all the short, and cheap, and easy ways of doing that whose difficulty is its honor—are just so many new obstacles in our already encumbered road. They will not make one of us happier or wiser.... They will only make us shallower in our understandings, colder in our hearts, and feebler in our wits.[16]

To Ruskin, one of the fundamental effects of mechanization had been the creation of a profound split between the mechanistic and the organic: stamped metals, often made using necessarily blunted details and worn dies, could never imitate the variety and casualness of nature. In *Modern Painters*, he more clearly defined the dichotomy between the natural and the mechanical as he complained that machine production necessitated making standardized copies of unique natural forms. Taking as an example the depiction of a "circling tendril, terminating in an ivy-leaf," he noted that in "vulgar design, the curves of the circling tendril would have been similar to each other, and might have been drawn by a machine, or by some mathematical formula. But in good design, all imitation by machinery is impossible. No curve is like another for an instant; no branch springs at an expected point. A cadence is observed," a cadence full of "its own change, its own surprises." The factor of surprise, the irregularity of natural forms, could only be created by handicraft, never by machines.[17]

In a century convinced of the educational and moral benefits of being exposed to examples of fine art, nineteenth-century American and British critics generally praised electrotyped reproductions of museum masterpieces, but British critics were often ambivalent about mass-produced industrial imitations, sometimes praising artful examples but also viewing the imitations as inherently degraded. Reflecting the century's mixed views toward reproductions and imitations, Christopher Dresser, nineteenth-century England's highly respected designer for industry and one of the country's most popular writers on design during the 1860s and 1870s, praised electrotypes of museum pieces in *Principles of Decorative Design* (1873) and recommended their purchase for study or sideboard, seeing them as "hardly inferior" and as having "equal value with the originals." But he dismissed electroplated imitations as "meretricious" and cast-iron copies made from newly designed molds as a "rough means of producing a result," concluding that although some examples of Berlin ornamental cast iron were "wonderfully good in their way" and even "to an extent artistic," ultimately no cast iron could compare with "works wrought by hammer and chisel."[18]

Rather than feeling alienated by factory-made imitations, however, American critics and commentators were often eager to embrace the new factory-made decorative arts and focus on their social benefits. Writing about the displays of art and industry at New York's Crystal Palace exhibition in 1853, Horace Greeley, editor of New York's *Herald Tribune*, insisted, "Our progress, in these modern times, then, consists in this, that we have democratized the means and appliances of a higher life, that we have spread, far and wide, the civilizing influence of Art." Through the greater availability of the decorative arts, Americans were "bringing, more and more, the masses of the people up to the aristocratic standards of taste and enjoyment."[19]

The fundamentally differing responses of American and British commentators were based not only on differing manufacturing methods but also on different historical experiences. America's generally sanguine view of machine-made decorative arts was in part a reflection of the country's success with standardized production methods. And in a country devoted to social progress and minimizing, theoretically, at least, the traditions of social class distinctions, American critics rejoiced in the possibility of democratizing the arts.[20]

The theme of splitting or separation that was seen in British critiques of the imitative arts was intrinsically associated with the new technologies themselves. As Cyril Stanley Smith has argued, when the ornament on many manufactured wares was made using a drop press that produced relief patterns on sheet metal, the "aesthetic qualities of the surface became divorced from the underlying substance, and decoration became independent of the body needed to support it."[21] Metal plating itself was a fundamentally divisive technology, for it placed a thin veneer over base material, creating a disparity between the expensive outer appearance and the underlying reality. British and American consumers, eager for the look of respectability and the trappings of wealth, welcomed these new veneers (including the new cast-iron building facades in America) even as British critics winced at the disparity between aristocratic-looking exterior and base interior, between lofty social pretensions and humble social origins.

William Morris, in his essay "The Arts and Crafts of To-Day," skillfully articulated many of the essential misgivings about the imitative arts, complaining that manufacturers, guided by the need to make profits, demanded that workers produce "makeshifts" of art, "commercial shams." The manufacturer not only forced these shams on the public but prevented the public "from getting the real thing." Prophetically voicing a complaint that would be made in postmodern critiques during the last decades of the twentieth century, Morris warned that the industrial imitations were replacing the originals: "the real thing presently ceases to be made after the makeshift has been once foisted onto the market."[22]

Living in the "age of makeshifts," Morris and the company he formed manufactured his own handcrafted wares, for as he argued, "we are the representatives of craftsmanship which has become extinct in the production of market wares."[23] The paradoxes and contradictions of Morris's practice have long been noted: though denouncing machinery, he acknowledged his own use of mechanized textile looms; rather than providing an alternative to mechanized production, his own handcrafted wares, with their elegantly abstracted or "conventionalized" natural motifs manufactured by skilled artists and artisans, turned out to be expensive—destined for the wealthy rather than a mass market. Yet his stylized aesthetic and his call for a renewed focus on craftsmanship had a marked impact on consumers in both Britain and America, who responded favorably to the Arts and Crafts movement. The aesthetic was also embraced by England's C. R. Ashbee and later Gustav Stickley in the United States, who sought to renounce the veneer in their quest for the real thing.

Morris, like Ruskin, assumed that there was no hope for reform as long as human beings were treated as machines. Reflecting the century's rhetoric of baseness and elevation, Ruskin in "The Nature of Gothic" had argued that workers engaged in mechanized production could only experience a type of reductionism: "All the energy of their spirits must be given to make cogs and compasses of themselves. All their attention and strength must go to the accomplishment of the mean act." However, when they were allowed to be human—to "begin to imagine, to think"—the "engine-turned precision is lost at once." Freed from the debasing impact of industrial labor, "out comes the whole majesty of him also; and we know the height of it only when we see the clouds settling upon him."[24]

William Morris, in "The Arts and Crafts of To-Day," saw the same dichotomy between the human and the machine. In an era in which manufacturers, the "captains of industry," were guided entirely by profit-making concerns, artisans, the once pleasure-filled craftsmen in touch with their own creations, had become depersonalized: the manufacturer's wares "must be made by instruments—as far as possible by means of instruments without desires or passions, by automatic machines, as we call them. Where that is not possible, and he has to use highly-drilled human beings instead of machines, it is essential to his success that they should imitate the passionless quality of machines as long as they are at work." Just as Ruskin concluded that "you can't have art where you have smoke," Morris asked rhetorically, "Need I say that from these human machines it is futile to look for art?"[25]

RECONCILING ART AND THE MACHINE IN INDUSTRIAL DESIGN

Though Morris and Ruskin focused on a theme of separation, seeing the artist-craftsman as alienated from the machine, efforts continued toward creating a more amiable sense of union between industry and art, and toward heightening the legitimacy of artists working for industry. In an age in which new versions of the imitative arts were being introduced as inexpensive alternatives to expensive, handcrafted originals made by skilled artisans, manufacturers and designers continued their efforts to gain respect for new machines and the factory-made decorative arts.

One of the central roles of the industrial designer and the "ornamentalist" was to help manufacturers integrate new industrial products into society. Whether stenciling sewing machine bodies with flowers or encasing cast-iron factory steam engines in neoclassical frames, nineteenth-century manufacturers aided by artists and designers were not only mirroring the design aesthetic that dominated the century but were also elevating the status of these products of innovative technologies. Their efforts to dignify manufactures with sometimes extravagant ornament received mixed reviews. Using the rhetoric of elevation and degradation so characteristic of the century, British architect and critic Matthew Digby Wyatt in his essay "Iron-Work and the Principles of Its Treatment" (1851), first published as an unsigned article in the *Art Journal*, praised Coalbrookdale's decorative cast-iron wares: "To the exertions made by the Coalbrookdale Company to elevate the character of the design of fancy casting too much praise cannot be given. The efforts they have made to elevate iron into a material for expressing the loftiest order of fine art, and the spirit with which they have enlisted the highest procurable sculpturesque ability redound to their credit."[26]

But ornamented manufactures and the imitative arts continued to generate a sense of crisis among those who lambasted the state of manufacturing design in Britain and America. During a century continuously searching for an appropriate industrial style, debates centered on finding the right "language," or grammar, for design. In 1846, the *Art-Union*, commenting on ornamented cast-iron stoves, criticized the practice of disguising poor design features with ornamentation: "confession indeed, is as honorable in art as in literature; it is as great a fault to disguise perplexity of thought under flourishes of ornament as to hide it under flourishes of language—the bombast of art like the bombast of letters betrays, instead of hiding, poverty of invention."[27]

Again using the metaphor of language, Matthew Digby Wyatt in his "Iron-Work" article encapsulated the terms of another important debate by contrasting design "utilitarians" with design "idealists." The utilitarians, the type who judged designs by a "narrow scale of human wants and necessities," produced designs "with more skill than taste," while the idealists were those who would "sacrifice comfort and convenience to ornament and effect." The idealist, Wyatt noted, tends to "speak a language with the grammar and component elements of which he is utterly unacquainted"; he applies with little understanding design bits and pieces of "Louis Quatorze shell-work or Gothic pannelling [sic], thinking that that will be quite *safe*."[28]

While William Morris fundamentally considered it "futile to look for art" from the hands of artisans who had themselves become machines, nineteenth-century educators were intent on improving the quality of industrial design, and bridging the gap between the idealists and utilitarians, by changing the education available to designers. Their aim was to create a closer sense of connection between art and industry, and to lend status and legitimacy to the nascent profession of industrial design.

In 1848 the *Art-Union* had argued for stronger ties between manufacturers and artists, and nearly thirty years later the American educator Walter Smith, in his book *Art Education, Scholastic and Industrial*, was still calling for a closer union: "The alliance of science and art in the factory and workshop is of great importance," he argued; yet improving the artistic quality of manufactures, heightening public taste, and elevating the designer's status proved to be no easy task. The sense of urgency for creating a closer union between art and manufacturing design was intensified by the impact of two major industrial design exhibitions, the Great Exhibition held in London in 1851 and the American Centennial Exhibition held in Philadelphia in 1876, exhibitions that became the very embodiment of national aspirations and pride. But to art critics and educators in both England and the United States, these and other international exhibitions also made all too evident the existing problems in manufacturing design.

Commenting on the Great Exhibition in London, England's *Journal of Design and Manufactures* in 1851 reprinted a disgruntled article published in the *Times* titled "Universal Infidelity in Principles of Design," which complained that "great atrocities of taste are committed," citing examples of manufactured lamps, candelabra, and chandeliers that were encrusted with entangled branches, cupids, and other mythic images. In an age that looked for fidelity to nature in ornamental imagery, the article also complained about art manufactures on which "animals and birds of all kinds are represented doing physical impossibilities." "On the whole," the article concluded, "the vein of art in connection with manufactures seems well-nigh exhausted all over Europe."[29]

Suggestive of attitudes associated with the influential British design reformer Henry Cole (editor of the *Journal of Design and Manufactures*) and his associates, the *Times* article also cited what it considered a chief source of the design problem: manufacturers and designers were too little concerned with utility: "All European nations at the present time begin manufacture with ornament and put utility in the background."[30]

Visits to European industrial exhibits brought a similar sense of dismay to Americans who questioned the aesthetic qualities of American manufactures. In a talk on industrial art education presented in Philadelphia in 1875, Walter Smith, who was professor of art education at the Boston Normal School of Art and state director of art education for Massachusetts, noted that a visitor to European industrial exhibitions would be "compelled in candor to admit that America has not taken that position in Art and Industry which she delights to take in respect to some other subjects." In a later talk before the Massachusetts Teachers' Association at Worcester in 1878, Smith again noted that in contrast to the art manufactures shown at the Paris Exposition of 1878, America "does not show well at such displays in comparison with many European countries."[31]

Probing the problem further, the United States Department of Interior's Bureau of Education, starting in 1885, issued a massive four-volume report on American education written by Isaac Edwards Clarke. Praising the "striking superiority" of American machinery shown at the Philadelphia Centennial, Clarke nevertheless noted that "to all observers who were not blinded by ignorance or prejudice," there was a "relative inferiority of Americans in most of the productive industries in which the fine arts enter as an important element." The works showing the "application of art to industry" were seen as "generally very inferior, both in excellence of design and in workmanship, to the articles of a similar class shown by exhibitors from other countries."[32]

In both England and the United States, the poor state of manufacturing design clearly seemed to point to the need for improved art education of designers and for an expansion of public art education to heighten standards of aesthetic taste and increase the demand for well-designed articles. Better design education would thus produce aesthetic, moral, and economic benefits, including heightened sales of factory-made versions of the decorative arts.

British and American educators assumed that designers needed improved, specialized training to create patterns for cast iron, textiles, carpeting, electroplate, and other manufactured articles, but they differed about what they considered the appropriate aesthetic for the new products. Nineteenth-century aestheticians and design theorists including Pugin, Ruskin, Cole, the designer Christopher Dresser, and British educator William Dyce (whose drawing book written in 1842 provided an early industrial design course manual for the British School of Design) all generally assumed that industrial designers should base their ornamental patterns on motifs drawn from nature, and more specifically from botany.

There was much debate, however, about whether the nature imagery should be mimetic or "conventionalized" (stylized imagery using flat patterns). Ruskin's own aesthetic favored illusionistic fidelity to nature, which led him to launch a caustic critique against those who would engage in conventionalized industrial design. In his lectures included in *The Two Paths* (1859), he complained that conventionalized designs used in manufacturers' machine-made decorative metalware in particular transformed the irregular, vital forms of nature into standardized mechanistic images. But British architect and designer Owen Jones, in his lectures and his influential pattern book *Grammar of Ornament* (1856), argued that the designer, or ornamentalist, should avoid the representational, illusionistic techniques used by painters to convey an accurate representation of nature. Instead, designers should strive for flat, stylized patterns as the only legitimate aesthetic for shaping the new industrial decorative arts.

Art critics and educational reformers also identified another fundamental problem that needed to be addressed—a problem that was, in some ways, endemic to the profession of industrial design itself: since the late medieval period in Europe, one of the distinguishing features of industrial design had been the separation of design from the process of making. Original designs or models produced by an individual designer were duplicated by artisans using craft methods and later by factory workers using machinery. In Italy and Germany in the early sixteenth century, craftsmen copied engravings in pattern books—further divorcing the original designer from those who duplicated the model.[33]

Eighteenth-century manufacturers obtained patterns from a variety of sources, including published pattern books and encyclopedias of ornament, and models that were based on antiquities or newly created by noted artists and employees working as craftsmen. During the second half of the century, designers in France had become independent specialists hired by manufacturers.[34] In England, Matthew Boulton purchased models and drawings for his dinnerware designs from well-known artists and architects such as the Adam brothers and John Flaxman; Josiah Wedgwood's designs included patterns not only by Joseph Wright but also by Flaxman and the artist George Stubbs.[35]

While nineteenth-century cast-iron manufacturers pirated patterns from the catalogs of other manufacturers, firms including England's Carron Company and Coalbrookdale commissioned models from prominent sculptors and designers. Coalbrookdale commissioned models by John Bell and purchased designs by Moyr-Smith and Christopher Dresser, among others. Elkington's electrotypes included specially commissioned designs by Antoine Vechte as well as French artist Pierre-Emile Jeannest and Léonard Morel-Ladeuil, both of whom were hired as employees.[36]

Problems arose, however, when manufacturers appropriated ancient, handcrafted models for mass production, or when designs were adapted from models created by artists ignorant of manufacturing processes, which then required revision by "technical designers" working for the manufacturer. As Cyril Stanley Smith has noted, metal casting by its very nature requires that the design include features that make it easy to separate the casting from the model. There were difficulties, however, when "the designer improperly copied features appropriate to earlier techniques" or was forced to blunt the details of more elaborate, hand-carved originals being used as models.[37]

Recognizing these problems, the *Art-Union* in 1848 noted, "As artists rarely trouble themselves with a study of the arts of reproduction, they are very liable to the error of introducing details into their works, which add little, if anything, to the general effect and beauty of the whole, but which interpose many serious obstacles to the process of modelling and casting."[38] A few years earlier, when the journal's editors toured the Midlands, they had found "no artist of any kind engaged at any one of the establishments in these great manufacturing towns." Occasionally the services of outside artists were used, but there were no members of an art staff employed in-house. More often, they found that "something was imported from abroad, and modified and adapted to suit the market." The results, as they reported, were "deplorable."[39]

The problem was still being discussed in 1882 by William Morris, who in his lecture "Technical Instruction" complained that designers were poorly trained and oblivious to manufacturing requirements. There are times, he noted, when a manufacturer "goes to a more dignified kind of artist, who, knowing nothing of the way in which the thing has to be done, produces a kind of puzzle for the manufacturer. The manufacturer, having paid for it, takes it away and does what he can with it . . . and so he chops the design up and adapts it to his purpose as well as he can; the design is spoilt, and when executed looks, not better, but worse, than the ordinary cut and dried trade design; so that the manufacturer has got no good by employing the great artist." Not surprisingly, Morris concluded that "the designer should be acquainted with the manufacture for which he is designing."[40]

In England, steps to create a specialized education for designers had begun in earnest in the 1830s, when the country was uneasy about the merit of its new manufactures, and as Quentin Bell has observed, there was "a widespread feeling that something must be done to educate artisans in art." A Select Committee was appointed in 1835 by the British Parliament "to enquire into the means of extending a knowledge of the arts and the principles of design among the people (especially the manufacturing population) of the country."[41] In the resulting 350-page report of 1835, James Morrison, member of Parliament, testified before the committee that "we have been very much superior to foreign countries in respect of the general manufacture, but greatly inferior in the art of design." London teachers were educating "people for painters and sculptors rather than as artists for manufacturers," he added, with the result that "though there are persons that are called designers, yet they have not been educated as such, and in point of fact they know little of the principles of art."[42]

As an outgrowth of the Select Committee's report, a national School of Design was established at Somerset House in 1837, and funds were made available to establish regional schools of design. The schools were patterned on French and Bavarian models, including French academies, which had long been training skilled artisans for industry.[43] In 1849, however, Henry Cole's *Journal of Design and Manufactures* reported the widely held view that the School of Design had hitherto been adjudged a "complete failure" in improving ornamental and decorative manufacturing in England, a failure that the journal attributed in part to the school's managers and directors, whom the journal considered "amateurs" lacking in knowledge about art.[44]

Changes in education followed. A Department of Practical Art was established in 1851 with Henry Cole as general superintendent and painter and educator Richard Redgrave as art adviser (in 1853 the department was superseded by the Department of Science and Art). Among the British regional schools established during the period was the School of Art founded in Coalbrookdale in 1856, which taught machine construction drawing, engineering drawing, practical mathematics, and applied mechanics. Its students were later employed in local industries, including china making and the art-castings department at Coalbrookdale.

The School of Design and Department of Practical Art headed by Cole became a recurrent source of controversy because of its educational policies, including Cole's commitment to providing commercial training for designers rather than focusing on a more general education in the fine arts. Cole and the department were both satirized in Charles Dickens's *Hard Times* in 1854, and John Ruskin in his preface to the *Laws of Fiesole* peevishly argued that "the very words 'School of Design' involve the profoundest of Art fallacies. Drawing may be taught by tutors: but Design only by Heaven; and to every scholar who thinks to sell his inspiration, Heaven refuses its help."[45] (This view was staunchly denied by subsequent educators in England and America and belied, in the early twentieth century, by the success of the design workshops of the Deutscher Werkbund and the Bauhaus.)

The need for educational development was also recognized in the United States. Isaac Edwards Clarke, in the 1885 government report on industrial art education, quoted from Walter Smith, who, echoing sentiments of the British Select Committee voiced fifty years earlier, had written, "It requires some courage to say deliberately that we are behind European nations in taste and skill." The problem, Clarke argued, could be remedied by a national system of art education comparable to the one in Britain: "it is only want of training in art industries, that is the cause of American inferiority in art products."[46]

Though there was no national system of industrial education, the training of designers in America had in fact already begun. As director of education in Massachusetts, Walter Smith had helped initiate a law passed in Boston in 1870 requiring that every child be taught drawing and that schools of industrial design be established in every city with a population over ten thousand. In 1872 a "free course of practical design for manufactures" had begun at the Massachusetts Institute of Technology in conjunction with the Lowell Institute which included courses in designing patterns for wallpaper and textiles. And the Department of the Interior's report on art education in 1897 included a large compendium of industrial design educational programs in cities throughout the United States.[47]

As another means of closing the gap between art and industry, art critics, designers, and educators urged that efforts be made to heighten the status of industrial designers. Recognizing the problem

40 Middletown Silver Plate Company, Middletown, Conn., "The Barge of Venus." Electroplate. *Art Journal* 38 (London), 1876.

in America, Walter Smith in his study *Art Education, Scholastic and Industrial* (1872) wrote that "in many places the idea prevails that, if a man or woman has not skill or imagination enough to become an artist, it is better to become a designer; in other words, that the weaker vessels of either sex, who cannot pass through the fine-art furnace, should be prepared, as coarser clay at a lower temperature, for the more ignoble occupation of pattern-drawers for the factories." Such a misconception, Smith claimed, "must inevitably result in impoverished and miserable design."[48]

England's *Art Journal* had already been keeping an eye on the status of industrial designers in both Europe and America, and in 1876 it published an admiring description of a plated sculptural centerpiece titled "The Barge of Venus" (fig. 40)—a cupid driving two swans pulling a gold-lined, shell-shaped barge, manufactured by the Middletown Silver Plate Company in Connecticut—noting, however, that the name of the designer was unknown. Admonishing American art manufacturers, the journal impressed upon them "the duty of rendering justice to the artists they employ," a practice the editors considered "sound policy," for "he must be an artist poor of soul who will not work better if he knows that fame is to be part of his reward."

The problem was also British, the journal acknowledged, for "we know the practice has been far too much neglected in this country."[49]

Efforts to raise the prestige of designers continued through the end of the century. In 1889, the *Art Journal* reported favorably on the situation in Germany, where industrial art benefited from royal patronage, state and municipal support to education, and "the open acknowledgment of individual talent, whether in the designer or the workman." The fine quality of German design stemmed from the country's willingness to place "the pursuit of Industrial Art as a profession on a level with Pictorial Art."[50] The quest to heighten the status of designers was also vigorously pursued by England's Christopher Dresser, who studied at the British School of Design from 1847 to 1854, specialized as a botanist, and received an honorary doctorate from the University of Jena in 1860 for his research and publications on botany. Emerging as one of Britain's most notable designers, he became known for his stylized floral Gothic cast-iron furniture designed for the Coalbrookdale Company in the 1870s, his ceramic designs, and the elegant simplicity of his Japanese-inspired electroplated teapots designed for British firms including James Dixon during the same decade.

Through his lectures and writing, Dresser became one of the most impassioned and articulate advocates for the emerging profession of industrial design as he attempted to lend honor, status, and nobility to industrial designers and the field of industrial design itself. In his essay "Ornamentation Considered as High Art," which appeared in 1871, he acknowledged that the art of design or ornamentation was frequently held in low esteem compared with the "higher" art of the pictorial artist and was "often slighted, and placed beneath its legitimate sisters." But Dresser remained ever hopeful and envisioned a Cinderella story that would reach its climax when "the despised sister has become the chosen one."[51]

Dresser's writing was filled with references to the ennobling role of designing and his chief preoccupation was with elevating the public's perception of ornamental design and "applied art." Design, in his view, lent value to otherwise worthless materials: as he wrote in the *Journal of the Society of Arts* in 1872, the "intrinsic value" of a manufactured article was in many cases "of less importance than the pattern which the article manufactured bears." As he noted with disappointment, however, the instructors at England's School of Design at South Kensington were pictorial artists rather than designers, so there was no one to instill in students a sense of design's "nobleness and greatness." "The time has come," he insisted, "when this state of things should be altered," so that "every facility should be offered to our young people for becoming great ornamentalists, and to our manufacturers for endowing their goods with an ennobling art." "Let ornament be treated as a fine art," he urged in his conclusion, "and as that which is ennobling and refining in its influence upon those who view it."[52]

Focusing directly on the professionalization of the industrial designer, Dresser presented a lecture on art schools in Philadelphia, where he had come to visit the Centennial Exhibition while en route to Japan. His lecture, published in 1877, presented a lyrical view of the designer: "the student of industrial art should receive such an education as will make him a scholar, a poet, a gentleman, and to an extent, a man of scientific knowledge." By "gentleman," he was quick to point out, he was not referring to social status but to "gentlemanly breeding," to "tenderness of perception and regard for the feelings of others."[53]

Dresser's continued aim was to elevate the status of industrial design to a respected profession by extending to designers monetary rewards and honors. He asked pointedly, "If the youth acquires the education which he certainly ought to have he will take his place amongst professional and scientific men," but will "the manufacturer pay such prices for designs as will remunerate the designer for his prolonged study?" To reward designers, Dresser added, universities should "confer degrees upon great ornamentalists, and certain honorary distinctions upon such manufacturers as produce ennobling works."[54]

As a reformer, Dresser was also out to improve the profession, for as he argued, "there are many pattern-drawers who are a disgrace to the profession, inasmuch as the works which they produce manifest only gross ignorance." What was most apt to "retard the progress of applied art" was "the production by designers of objects which are inconvenient to use." Offering a vision of aesthetic integration, he insisted: "That which is most beautiful may be perfectly useful."[55]

In *Principles of Decorative Design* (1873), Dresser included design sketches illustrating functional applications of handles and spouts to teapots and urged designers to make ornament an integral element of design, paralleling the form instead of separating object and ornamentation. To create relief patterns on a vase, he advised, "follow the line of the vase" rather than produce "irregular projections from it."⁵⁶ Dresser's own designs, particularly his electroplated teapots with their slanted ebony handles and ornament-free, spare simplicity, offered a remarkable linking of aesthetic elegance and functional design. Through his exemplary vision, he revealed that designers could make a difference by imaginatively reconciling art and the machine, creating a sense of clarity and coherence after a century of fractured change.

Chapter 4

The Anxiety of Imitation
Electrometallurgy and the Imitative Arts

Reflecting the nineteenth century's fascination with factory-made reproductions and imitations, British and American manufacturers introduced their new electroplated and electrotyped copies of costly silverware, vases, sculptures, and other ornamental wares to a middle-class market eager for the look of luxury. Official recognition and acceptance of these new technologies were clearly evident at the century's great international industrial exhibitions, where decorative examples were prominently displayed and publicized. At the exhibitions, electroplating and electrotyping were presented as yet another triumph in an age of progress, and the new technologies were used by manufacturers to fashion large-scale emblems of the era's optimism and pride in technological achievement.

At London's Great Exhibition of 1851, George Elkington proudly presented his firm's four-foot-high electroplated vase, which, as the *Art Journal* noted, represented the triumph of science and the industrial arts. This showpiece earned Elkington the exhibition's Council Medal, its award of highest distinction (fig. 41).[1] The vase was surrounded by four statuettes of Newton, Bacon, Shakespeare, and James Watt commemorating astronomy, philosophy, poetry, and mechanics, and at the base were allegorical figures of War, Rebellion, Hatred, and Revenge, considered to be "overthrown and chained" by the beneficial impact of art and science. At the top was a sculptural figure of Prince Albert, the exhibition's mentor and champion of the industrial arts.

41 Elkington, Mason & Co., "The Triumph of Science and the Industrial Arts." Electroplate, designed by William Beattie. *Art-Journal Illustrated Catalogue of the Great Exhibition of Industry of All Nations*, 1851.

42 Reed & Barton, "The Progress Vase" (1876). Electroplate. *Art Journal* 38 (London), 1876.

Not to be outdone, American electroplate manufacturers exhibited their own icons of national progress at the Philadelphia Centennial Exhibition held in 1876. Winning a commendation from the exhibition's judges was the electroplated "Progress Vase" manufactured by Reed & Barton, a large ceremonial trophy designed by William Beattie, the designer of Elkington's vase, who had come to Massachusetts from England to work for Reed & Barton in 1875 (fig. 42).

Intended as an allegory, the four-and-a-half-foot-tall vase represented the progress of America from the fifteenth to the nineteenth centuries, or, as the educator Walter Smith put it, "the progress of America from savage to civilized life." The vase's themes of war and peace echoed the themes on Elkington's version, now reinterpreted for the New World. On the left were figures of fifteenth-century Aztec Indians learning the art of war, and on the right were the god Mercury and the figure of Columbia walking with an allegorical Plenty carrying a cornucopia signifying agricultural prosperity. A bas-relief presented the landing of Columbus carrying olive branch and dove, and topping the trophy was the figure of Liberty carrying a scroll inscribed with the word "progress."[2]

The triumphant display of these vases and trophies appeared amidst a larger, century-long debate about the design, social impact, and legitimacy of the new die-stamped, plated-metal reproductions and imitations. While Americans tended to welcome the imitative arts and the new developments in electrometallurgy, British critics, though sometimes praising the idea of bringing decorative wares within the economic reach of many, winced at the social and aesthetic pretensions of the new imitative arts—imitations that threatened to blur social class distinctions and challenge the privileged position of the original work of art.

Reviewing America's industrial products on display at the large exhibition of 1853 held at the Crystal Palace in New York, Horace Greeley, editor of the New York *Herald Tribune*, defended the manufacturing of art metalwares, arguing that although not the equal of handcrafted Renaissance masterpieces, America's factory-made metalwares had a valuable democratizing impact: "Our vases and cups may not be more exquisitely wrought than the vases and cups of Benvenuto Cellini, but they are wrought, not like his, for Popes and Emperors, but for Smith and Jones."[3] But to Britain's John Ruskin, the value of machine-made metalware was thoroughly undermined by its imitative nature. As he wrote in *The Seven Lamps of Architecture*, "the substitution of cast or machine work for that of the hand" was deceitful, and the deliberate creation of deceitful ornaments, he concluded sternly, was "an imposition, a vulgarity, an impertinence, and a sin."[4]

TECHNOLOGICAL DEVELOPMENT

The common focus of these mixed critical reactions was the growth in new casting and plating processes, as well as new means for die-stamping and embossing metal with ornamental surfaces. Electrotyping developed through the almost simultaneous, independent discoveries by Russian physicist Moritz Hermann von Jacobi, British scientist Thomas Spencer, and British inventor C. J. Jordan in 1837 of processes in which engraved plates, immersed in chemical solutions, were subjected to electrical current that left a copper deposit which, when removed from the original plates, formed a relief copy of the engravings. The practical applications of electrotyping were recognized almost immediately, including its uses for reproducing line engravings and copying artworks. In later applications, museum pieces as well as art specially commissioned by manufacturers were reproduced through electrochemistry by first creating a mold of the original in plaster of Paris, wax, or clay. These nonmetallic molds were rubbed with graphite or black lead to render their surfaces conductive, and they were then hung in a conducting solution, with a copper sheet also suspended in the solution. A flow of electric current gradually transferred copper from the copper sheet through the solution and deposited it on the conducting surface of the mold. The mold was then removed, leaving a copper shell that was an exact duplicate of the original. This electrotyped copy was then coated with gold or silver using an electroplating process.[5]

Spencer's work, published in 1839, was soon popularized in America, and his paper "Instructions for the Multiplication of Works of Art in Metal by Voltaic Electricity" was reprinted in Benjamin Silliman's *American Journal of Science* in 1840. In 1841, the work of both Jacobi and Spencer was included in Alfred Smee's *Elements of Electro-Metallurgy*, published in London and widely read in the United States. Within a decade, electrotyped art reproductions were being seen and admired by huge audiences, such as those at London's Great Exhibition of 1851 and New York's 1853 Crystal Palace exhibition.

Electrotyped replicas of well-known artworks became part of the collections of major art museums, including London's South Kensington Museum, later renamed the Victoria and Albert Museum, which exhibited electrotyped copies of works such as *The Sacrifice of Isaac* by the Italian Renaissance sculptor and architect Filippo Brunelleschi, the bronze sculptural relief entered in the competition for the doors of the Florence Baptistery. The museum also commissioned Elkington to make electrotyped reproductions of its own holdings, and Elkington's firm itself commissioned noted contemporary sculptors such as British artist John Bell to create original sculptures for electrotype reproduction.[6]

Changes in die-stamping, embossing, and plating methods were also underway. Dies had long been used to stamp patterns on metal pieces, but beginning in the 1830s, improved machine tools and the increased number of British patents for molding, stamping, and embossing metals helped manufacturers produce relief designs on sheet metal, though the early dies were often poorly cut or became worn after prolonged use, producing a large number of crudely cut and stamped machine-ornamented products that only added to critics' dismay.[7]

Introducing a new process for plating, British manufacturer George Elkington and his cousin Henry took out a patent for electroplating in 1840 using the previously known technique of electrodeposition to plate silver and gold onto base metals. Metal plating itself, though, had a long tradition. Gold since early times had been spread in thin sheets and hand-hammered or pressed against three-dimensional dies of wood, stone, and metal, a process described by the artist, craftsman, and Benedictine monk Theophilus in his manuscript treatise *De diversis artibus* (1125 A.D.). Also an ancient tradition was the plating or resurfacing of base materials with a metal coating in order to produce a more expensive appearance. Virtually every civilization has covered cheap materials with gold and silver; silverplating was produced in ancient cultures by rubbing base metals with a mixture of silver chloride and chalk, or rubbing with gold. Theophilus described the soldering of gold on silver and silver on copper, and medieval practitioners applied thin gold sheet to finished metal objects.[8]

During the eighteenth century, Sheffield silverplating or fusion plating employed a technology said to have been first developed by England's Thomas Boulsover in the 1740s. A sheet of silver was fused through heat to one or two sides of a copper ingot, which was then rolled into sheets from which articles were stamped out (in silverplating, the plating took place before the object was fashioned; in electroplating, the object was plated after it was made). The base as well as the neck of candlesticks were often die-stamped and then soldered together. Skilled craftspeople shaped and assembled the components, and to speed production, ornamental details were also die-stamped. Sheffield became widely known for its silverplate, though the term "Sheffield plate" referred not only to silverplate made in Sheffield but also to that made by Birmingham manufacturers, including Matthew Boulton's Soho firm. Sheffield plate was later aided by the introduction of britannia metal (an alloy of tin, 4–8% antimony, and 1–2% zinc) as a base that could easily be stamped or cast from a mold and plated with silver.

In their efforts to imitate silver, early silverplate manufacturers stamped their wares with simulated versions of the hallmarks used on genuine silver decorative wares. Improvements in the silverplating process increased the illusion of solid silver: the copper bottoms of trays were coated with tin, and after 1785, silver coatings were placed on edges where the copper was likely to wear through. The use of the white-metal alloy britannia also helped to alleviate this problem by making the wear less obvious. During the period from 1770 to 1800, Sheffield plate often rivaled contemporary cast and chased silver in appearance through the use of steel dies to create fine ornamental detailing, but later, when manufacturers started using stamped silver foil filled with solder, the silver surface, when worn, revealed the copper base. By 1840, the introduction of relatively inexpensive electroplated silver signaled the waning of silverplate's popularity.[9]

Challenging Sheffield plate's hundred-year domination of the commercial market, George and Henry Elkington took out their new patent for electroplating using the electrolyte potassium cyanide to produce a firmer plating than had been manufactured previously. The process of electroplating often began with the manufacture of an object such as a teapot made of copper, brass, britannia, or the more durable alloy known as German silver or nickel silver (previously imported from China and known as paktong, the alloy was made of nickel, copper, and zinc). Steam-powered die-stamping techniques were used to produce individual components including the spout, which were then soldered together and polished. Embossed and engraved ornamental patterns were then added, and the object was brought to a plating room.[10]

In an article on Elkington's factory on Newhall Street in Birmingham, London's *Illustrated Exhibitor* in 1852 described the firm's "stupendous machine," a steam-powered generator, for the application of electric current, a mechanism capable of depositing fifty ounces of silver per hour. The process was similar to electrotyping, but did not use a mold. The object to be plated was placed in a vat filled with a chemical solution and bars of pure silver or gold. An electrical current passing through the solution dissolved the silver from the bars and deposited it on the immersed articles. The final steps were burnishing by women workers who used small steel burnishers, water, and a bit of rouge to give the objects a polished, lustrous appearance.[11]

Electroplating was commercially successful in both England and America, and was used to manufacture a wide variety of products from silver waistcoat buttons to ornamental vases, butter dishes, and lamp bases. In Victorian England, which favored ornamental motifs taken from nature, it was not considered bizarre to coat actual butterflies, acorns, birds, lizards, and tortoises with black lead or phosphorus to make them conductive, dip them into chemical solutions, and electroplate them with silver. The technique became hugely popular, and Elkington manufactured kits for home use.[12]

The popularity of electroplating in Britain was in full force in 1876 when it was reported that Sheffield plate had been almost totally superseded by electroplating and that Elkington's firm remained the leading producer.[13] American electroplating became equally successful after the country's manufacturers, sometimes aided by British platers, began electroplate experiments in the 1840s, with Henry Reed of Reed & Barton recording his own early experiments in his notebook of 1850.[14] By the 1860s, plated decorative wares—which were produced through the electroplating process but variously referred to as electroplate, electrosilverplate, and silverplate—were widely available in the United States, produced by manufacturers including Rogers Bros., Reed & Barton in Taunton, Massachusetts, the Gorham Company in Providence, Rhode Island, and the Meriden Britannia Company in Meriden, Connecticut.[15]

In an early article promoting the new American silverplate, *Harper's New Monthly Magazine* reported in 1868 that although five years before "all the really serviceable plated ware" had been imported from Sheffield and Birmingham, now only the cheaper forms of electroplate were imported due to the success of American manufacturers such as Gorham (which had begun its productions relatively late, in 1865).[16]

In their published catalogs, American electroplate manufacturers including the Connecticut firm Simpson, Hall and Miller in 1879 presented a full range of the decorative wares popular during the nineteenth century, such as candelabra, lidded soup tureens, calling-card receivers, and the omnipresent six-piece tea service consisting of three pots, a sugar bowl, creamer, and waste bowl. These electroplated wares were significantly less expensive than solid silver. A solid silver tea service by Gorham cost from four hundred to twelve hundred dollars during the 1850s; ten years later the Meriden Britannia Company, in its electroplate catalog of 1867, offered a plain, six-piece electroplated tea service for thirty-six dollars or an ornamented version in its "charter oak" pattern for fifty dollars.[17]

The styles chosen for the new decorative electroplating in both England and America reflected the century's fascination with what was then termed "naturalism" and the era's tendency toward eclectic amalgams of historicized styles. In England, the 1840s witnessed the popularity of styles based on natural or organic forms such as an electroplated fruit dish in the form of a vine leaf. Grape bunches, plant leaves, and small sculpted birds and animals were often entwined around or mounted on a teapot handle, spout, and finial.

Forms copied from nature were often merged with stylistic references to neoclassical, Renaissance, and rococo imagery. From the 1850s to the 1870s in England, the newest fashion was the "Etruscan" style based on the vessel shapes and ornamentation of classical Greece and Rome, seen in classicized electroplated coffeepots showing simplicity, symmetrical balance, and controlled formality.

While initially copying designs from British manufacturers, American electroplate manufacturers increasingly took out their own patents. Reed & Barton took out its first design patent in 1858 for an electroplated tea set.[18] American electroplate wares were also replete with incised geometric patterns, machine-stamped ornaments, and molded ornamental leaves, flowers, and vines as well as neoclassical motifs, often inventively combined with small sculptural figures. Reed & Barton in 1849 produced a chastely elegant teapot in the shape of a neoclassical urn, but by the 1860s the company was producing fanciful wares such as an oval cake basket adorned with miniature classical busts and a small dog's head on the handle. Joining history and myth, Meriden Britannia Company produced a four-gallon plated wine cooler and punch bowl decorated with a water nymph and small figures of crouching Civil War soldiers with rifles pointed (fig. 43).

THE CRITICAL RESPONSE

Predictably, the success of technologies that could create credible copies of costly decorative objects caused elation in engineering circles and feelings of despair among a number of British critics. As befitting a profession devoted to solving technical problems, engineers delighted in developing methods to produce metal reproductions and imitations as efficiently as possible. Fears about the aesthetic impact of creating copies of original works of art were only rarely considered. But British social commentators and art critics voiced serious concerns about the industrial imitations. Their argu-

43 Meriden Britannia Co., Plated wine cooler and punch bowl (1886). Courtesy of Christie's, New York.

ments often focused on the aesthetic appearance of manufactured art metalwares as they searched for the right language and grammar of ornament, but behind their aesthetic debates often lay unresolved, gnawing questions about the moral and social legitimacy of these new industrial imitations.

British critics in the 1840s were particularly incensed at die-stamped, plated, and cast-iron art manufactures, arguing that they violated canons of aesthetic taste, handicraft, and Victorian notions of naturalism. Surveying contemporary examples of machine-stamped clocks designed as ersatz Gothic cathedrals and cast-iron stove grates imitating the look of turreted castles, A. W. Pugin, the articulate champion of Gothic revival, presented in *The True Principles of Pointed or Christian Architecture* (1841) a withering indictment of contemporary metal manufactures, dismissing the cities of Sheffield and Birmingham as "those inexhaustible mines of bad taste."[19]

But the most articulate and influential skeptic was John Ruskin, who in *The Seven Lamps of Architecture* enunciated two of his central criticisms against industrial imitations: they were vulgar and immoral. Although there was historic precedent for factory-produced decorative arts and plated base metals, nineteenth-century critics including Ruskin feared the increasing dominance of the mechanical over the natural. In a stirring indictment of the industrial age, he again championed the organic over the mechanistic, setting the terms for the arguments that would inform the writing of the British design reformers including Henry Cole, Owen Jones, William Morris, and the British Arts and Crafts movement:

> There is dreaming enough and earthiness enough, and sensuality enough in human existence, without our turning the few glowing moments of it into mechanism; and since our life must at the best be but a vapour that appears for a little time and then vanishes away, let it at least appear as a cloud in the height of Heaven, not as the thick darkness that broods over the blast of the Furnace and the rolling of the Wheel.[20]

But Ruskin's eloquently voiced objections to machine-stamped designs went beyond the split he perceived between the natural and mechanical. In his essay "The Lamp of Truth" in *The Seven Lamps of Architecture*, he made vividly clear the moral framework that so firmly shaped his design aesthetic as he roundly condemned the vulgarity and inherent deceitfulness of reproductions imitating handmade work—what he called "these vulgar and cheap substitutes for real decoration."[21]

Ruskin's view of vulgar imitations clearly reflected his own work ethic: machine-made imitations allowed for shortcuts, the circumvention of human labor, a practice he had sharply denounced as "cheap and easy." The true delight, worth, and preciousness of handmade ornament lay in the viewer's ability to trace the path of the artisan's "intents, and trials," and the time spent in creation.[22]

Though British critics' complaints about the imitative arts seemed to be mostly concerned with matters of aesthetic honesty and morality, their language also revealed their concern with social status and privilege. In a social climate sensitive to nuances of class distinctions, these critics often couched their complaints in terms of social overreaching. Pugin in *The True Principles of Pointed or Christian Architecture* voiced his anxiety that decorative cast iron would make the trappings of status available to those in the lower classes: "Cheap deceptions of magnificence encourage persons to assume a semblance of decoration far beyond either their means or their station, and it is to this cause we may assign all the mockery of splendour which pervades even the dwellings of the lower classes of society."[23]

Charles Dickens in his novel *Our Mutual Friend* (1865) turned his gift of keen satiric observation to vulgarity and social pretension. Taking a mordantly comic view of the Podsnap family, he pictured their extravagantly ornamented tableware as a version of aesthetic eczema:

> Hideous solidity was the characteristic of the Podsnap plate. Everything was made to look as heavy as it could, and to take up as much room as possible. Everything said boastfully, "Here you have as much of me in my ugliness as if I were only lead; but I am so many ounces of precious metal worth so much an ounce;—wouldn't you like to melt me down? A corpulent straddling epergne, blotched all over as if it had broken out in an eruption rather than been ornamented, delivered this address from an unsightly silver platform in the center of the table.[24]

Other nineteenth-century British critics often wrestled with their own ambivalence about the plated metals and imitations. Articulating its aesthetic credo, the *Journal of Design and Manufactures* in 1851 took a firm stand toward machine-stamped metalwares: "Our readers well know that we have no toleration for SHAMS and IMITATIONS of any kind, as we consider them destructive of good art."[25] But paying tribute to the technology, the *Journal* also made a point of praising an electroplated tea and coffee set manufactured by Dixon of Sheffield for its "simple and sensible forms" and proudly viewed electroplating itself as a source of national pride: "In plated metalwork, England may expect to stand preeminent in the Exhibition of '51."[26]

The heavyweight excesses of contemporary plated metals were targeted by the British tastemaker Charles Eastlake, who in his *Hints on Household Taste* (1868) pointedly denounced "deplorable examples of taste," seeing in modern design only misguided efforts by manufacturers to achieve status through imitation. He scornfully dismissed contemporary plated designs "dignified by fine names such as 'Albert,' the 'Brunswick,' the 'Rose,' the 'Lily,' " finding "no more *art* in their design than there is in that of a modern bed-post." He warned consumers that if they continued to "insist on the perpetuation of pretentious shams" and preferred "a cheap and tawdry effect to legitimate and straightforward manufacture," then "no reform can possibly be expected."[27]

While British critics joined Dickens in taking potshots at the Podsnaps' silver, electrotyped copies of museum masterpieces won a warm reception from British and American scientists and engineers as well as art critics. England's *Art-Union* had, however, initially taken a more cautious view when it introduced its readers to the new method for producing exact copies of engraved plates in 1841. Revealing some ambivalence, the journal insisted that the invention was a "valuable one" and acknowledged electrotyping's usefulness in duplicating tens of thousands of engravings, but worried that it would also "depreciate the value of work; for nothing has yet been discovered to equal manual application in engraving."[28]

But ten years later, electrotyped reproductions of museum originals were effusively praised by those intent on increasing the public's exposure to fine art. Ralph Nicholson Wornum, in his essay published in the *Art Journal*'s illustrated catalog of the Great Exhibition, saw the "galvano-plastic art" as "destined eventually to perform a great part in the dissemination of taste, and in general education."[29]

It was this potential for elevating public taste that made electrotypes as well as decorative electroplating particularly seductive to America's cultural commentators. American scientists and engineers were quick to praise electrotyped facsimiles. In 1852, *Appleton's Mechanics' Magazine and Engineers' Journal* edited by Benjamin Silliman, an American scientist, devoted several pages to the application of electrometallurgy to sculptural bas-reliefs and included excerpts from Smee's text on electrometallurgy, then in its third edition.

But it was less the technical achievement than the possibilities for social improvement that attracted American reviewers to electrotyping, electroplating, and the imitative arts. In a country still striving to establish its own cultural identity and forge a national aesthetic style, the new technologies provided a welcome means of heightening the nation's own sense of its stature and pride.

Sensitive to the need for improvement, American writers of the period were often defensive about the appearance of their country's manufacturing designs. In *Art and Industry* (1853), his review of the Crystal Palace exhibition, Horace Greeley argued that Americans who visited the exhibition would see European "models of workmanship which are superior," adding that "in the finer characteristics of manufacture and art, we have yet a vast deal to learn." While praising America's technological achievements, he emphasized the country's continuing aesthetic inferiority: "stupendous as our advances have been in railroads, steamboats, canals," and "adroit, ingenious and energetic as we have shown ourselves," Americans still have "few fabrics equal to those of Manchester, few wares equal to those of Birmingham and Sheffield." "These are the articles that we ought to have," he concluded, "to relieve us from dependence on other nations, to refine our taste, and to enable the ornamental and elegant appliances of our life to keep pace with our external development."[30]

As the reference to taste suggests, Americans at midcentury voiced their insecurity about the appearance of their industrial products and the level of their aesthetic sophistication. Greeley clearly assumed that nouveaux riches Americans needed to be schooled in the subtleties of taste. There is a typically American preoccupation with achieving social and aesthetic legitimacy in his comments on wealth: "Mere wealth, without the refinements of wealth—barbaric ostentation, prodigal display, extravagant self-indulgence—can only corrupt morals and degrade character. But the cultivation of the finer arts redeems society from its grossness, spreads an unconscious moderation and charm about it, . . . elevates our ideals, and imparts a sense of serene enjoyment to all social relations."[31]

While praising America's "common people" for their superior intelligence and "sterling virtues" compared with their counterparts in other countries, Greeley still described ordinary Americans as aesthetically inferior, as "almost immeasurably behind them in polished and gentle manners, and the love of Music, Painting, Statuary, and all the more refining social pleasures."[32] Amidst this unsureness about America's social status, American commentators welcomed the availability of the imitative arts as a way to elevate public taste. John Leander Bishop in his *History of American Manufactures* (1868) praised electroplate manufacturer Reed & Barton, "where chaste designs are multiplied and wares rivalling those of the jeweller and silversmith are made and sold at prices accessible by all," helping Americans to become "educated in taste and love of the beautiful."[33]

Benjamin Silliman in *The World of Science, Art, and Industry* (1854) joined in the praise for the imitative arts and embraced the romantic belief—shared with British design reformers—in the "civilizing influence of art." While the French scientist Réaumur had fretted about the leveling aesthetic effect of having too many beautiful objects available through reproductions, Silliman argued that if art in the form of metal casts, electroplated copies, and ornamented furniture were "within the reach of the mechanic and tradesman as well as the opulent and noble," and "if the beautiful were daily placed before us, surely our social life could not fail to be ameliorated and exalted by its silent eloquence."[34]

While British critics were busy scoffing at the pretension and dishonesty of the factory-made imitations, Americans viewed the new imitative arts as socially beneficial rather than pretentious: items such as electroplated vases and candlesticks, they asserted again and again, could not only bolster America's cultural standing but would also be great equalizers, bringing decorative wares formerly the province of the privileged into the homes of many Americans. Horace Greeley in his *Tribune* articles reviewing New York's 1853 exhibition indulged in his usual expansive rhetoric as he praised the country's science discoveries and their applications to the arts for helping fulfill America's aesthetic and democratic objectives: "we have *universalized* all the beautiful and glorious results of industry and skill, we have made them a common possession of the people; and given to Society at large—to the meanest member of it—the enjoyments, the luxury, the elegance, which in former times were the exclusive privilege of kings and nobles." America's progress, he argued, consisted in having "spread, far and wide, the civilizing influence of Art," in bringing "the masses of the people up to the aristocratic standard of taste and enjoyment, and so diffusing the influence of splendor and grace over all minds."[35]

The new technologies of electrotyping and electroplating thus had quickly become connected with the country's wish to establish the superiority of its industrial products, fulfill its longing for cultural sophistication, and satisfy its need to lend social imprimatur to an upwardly mobile middle class. Electroplating, a technology in which base metals were ornamented to imitate the look of wealth and luxury, was particularly popular among a rising middle class eager to acquire its own patina of affluence.

In a country with a middle class eager to achieve cachet—and to have a life style indistinguishable from that of the wealthy—it was not surprising that art commentators increasingly began arguing that the factory-made imitations and reproductions were indistinguishable from the originals. Reporting on the "Gorham ware" electroplating at the Philadelphia Centennial, *Frank Leslie's Historical Register* wrote that "the resemblance of this to genuine silver is so close, that marks have to be resorted to for indication."[36] By 1876, the *Art Journal* concluded confidently that the high aesthetic, educational, and moral value of electrotyping had virtually eliminated the inherent differences between the reproduction and the original. In this endorsement of the technology, the prestige and integrity of the original, handcrafted work was less of an issue, for the factory-made electrotyped copy had achieved a legitimacy and status all its own. Reviewing an exhibit of Elkington's decorative electrotypes that was on display at the Philadelphia Centennial Exhibition in 1876, the *Journal* approvingly wrote:

> For all the purposes of Art—to give pleasure, to refine taste, to convey instruction—the electrotype is quite as good as the original in costly metals of gold or silver; indeed, it may be a question which would be preferred; there is no sort of difference except in the intrinsic value of the material, and that, as compared with the art lavished upon it, is very little.[37]

Harper's New Monthly Magazine, in an article praising American metal plate manufacturers in 1868, provided a revealing portrait of its middle-class readers' preoccupations with class and status. Filled with class references, the article's language referred to nobility and baseness, civility and animalism, the high and low. In the glittery world created for *Harper's* readers, a dining table covered with silver and silverplating became a cultural sign of aristocratic living: for "nothing conveys a more vivid impression of royal magnificence and imperial state."[38]

Appealing to mid-Victorian values, *Harper's* assured its middle-class readers that the use of silver for dining would help them to establish their sophistication, prove their moral rectitude, and achieve a comfortable distance from baseness: "It is a duty we owe ourselves and one another to glorify and refine eating and drinking, so as to place an infinite distance between ourselves and the brutes, even at the moment when we are enjoying a pleasure which we have in common with them." The choice was clear: an increase in wealth would bring either "tasteless and barbaric pomp" or "higher refinement, better education, and more elegant modes of living. We shall either spend our surplus money in pleasures that ennoble or in pleasures that debase."[39]

Like the middle class eager to imitate the long-established elite, nineteenth-century electroplate manufacturers were eager to present their newly arrived technology as an established member of the design community. Their products, they argued, were the genuine article, superior to counterfeit versions. Their quest for legitimacy gave rise to an anomaly of the imitative arts in nineteenth-century America: consumers were warned not to buy imitations of imitations (an argument that must have reverberated with a middle-class market aware, at some level, of its own counterfeiting of a lofty life style).

Praising its own "fine electro-plate," the Gorham Company in an advertisement published in 1868 warned of "English Imitations" as well as imitations by other American manufacturers. By going to reputable dealers, consumers could "avoid counterfeits by noting the distinctive trade-mark on every article."[40] In its article, *Harper's* also made a pointed distinction between the real and the fake, between authentic plate and fraudulent imitations. The magazine even saw the potential for deception and falsehood in the earliest stages of electroplate manufacture, for in the processing of silver to be used for plating, manufacturers first needed to "pick out the counterfeit coins" that were in circulation before the Civil War.[41]

More class overtones were present when *Harper's* told its middle-class readers that the best grade of plate had a suitably thick veneer. Introducing the electroplating process, including the deposition of silver in solution, the magazine warned, "Here is the opportunity for deception." Genuine or legitimate plate, it argued, generally required three to seven hours for the proper thickness of silver to be deposited, but inferior imitations might receive a coating in only three minutes and, with polishing, acquire "the highest degree of brilliancy." After a few days of use, though, "the brilliancy vanishes," exposing the inferiority of the imitation.[42]

As another anomaly, Americans who had argued in favor of democratizing the decorative arts began to insist on status distinctions between different grades of electroplating, again echoing the middle class's own status anxieties. *Harper's* distinguished between superior and inferior imitations, between "cheap," trashy versions of electroplating and much more elegant and expensive versions that could pass for an original. Praising the Gorham Company, the magazine insisted "they do not offer the public that insensate trash, which looks more radiant than plate itself for a few weeks or months, and then rapidly fades away into shabbiest brass or dents into manifest pewter." In making class distinctions between levels of plating, in establishing the gentility of "genuine" plate, *Harper's* editors were not

44 Hukin and Heath (Birmingham and London), designed by Christopher Dresser, Electroplated teapot and stand with ebony handle (1878). By courtesy of the Board of Trustees of the Victoria & Albert Museum.

only validating the new technology but also reaffirming their readers' own sense of gentility, of status achieved.

In England, amidst criticisms of plating as a pretentious, overwrought technology, electroplating nevertheless achieved recognition, as in America, for its potential to elevate public taste. It also achieved status through the achievements of individual designers, particularly Christopher Dresser, whose designs for plated metals, with their radical simplifications and striking elegance, helped heighten the prestige of British manufacturing design. Acknowledging the early influence of Owen Jones, whose *Grammar of Ornament* (1856) popularized flat, abstracted patterns adapted to utility, Dresser evolved his own industrial design aesthetic which had a remarkably restrained and elegant functional simplicity.[43]

Trained as a botanist, he worked first as a botany professor and lecturer before designing for major British electroplate manufacturers, including Elkington starting in 1875, Hukin and Heath starting in 1878 (through 1881), and James Dixon and Sons in Sheffield starting in 1879.[44] With Dresser working as design adviser to Hukin and Heath, the firm's plated products were praised by the *Art Journal* in 1879 for "singularly excellent Artworks, vast improvements of the 'have beens' of earlier times."[45]

The spare simplicity of Dresser's designs reflected his fascination with a Japanese aesthetic, and he made a trip to Japan in 1877, where he began writing *Japan: Its Architecture, Art and Art Manufactures*, published in London in 1882. Dresser's electroplated teapot with triangular stand and ebony handle designed for Hukin and Heath in 1878 and his electroplated teapot made for James Dixon and Sons of Sheffield in 1879 echoed Japanese design motifs and also countered the nineteenth century's pervasive infatuation with historicized imitations. His abstracted designs were a dramatic departure from the ornate rococo and neoclassical electroplate patterns being produced by the same manufacturers at the time (fig. 44). Dresser's designs were commercially successful, and by 1885 plagiarized imitations of his elegant simplicity were being manufactured by firms including James Dixon and Sons, which produced its own glass decanter mounted on electroplate with a wooden handle.

As Dresser's maverick electroplate designs suggest, there were some movements afoot for more chastened designs. In the early decades of the next century, the excesses of the cheeky, upstart imitative arts that had so unnerved British critics and lured Americans were being tempered by an emerging industrial aesthetic that made a doctrine of unornamented simplicity and functional design. These designs represented industrial societies convinced enough of the machine's status and legitimacy to be able to forgo the borrowed prestige of historical imitation. Modernism would engage in its own imitation: not the imitation of a handcrafted look, not the imitation of styles of the past, but the look of clean, spare machine forms.

Chapter 5

The Struggle for Legitimacy
Cast Iron

Of all the nineteenth-century technologies used to create reproductions and imitations of the decorative arts and architectural ornament, cast iron proved to be one of the more controversial. During the early decades of the century, manufacturers in both Europe and the United States challenged aesthetic and social traditions by producing not only cast-iron architectural ornament, a practice that had begun in the late eighteenth century, but also sculpture, decorative objects such as vases and candlesticks, and home furnishings. Following the lead of Prussia, the Coalbrookdale Company in Shropshire, Britain's leading iron manufacturer, issued catalogs starting in 1846 that advertised ornamented cast-iron stove grates as well as neoclassical vases, vine-patterned garden chairs, and cast-iron hall furniture designed in the stylized medieval patterns popular during the period. At midcentury, New York manufacturers James Bogardus and Daniel Badger dramatically challenged architectural traditions with their ornamented cast-iron building facades that imitated the look of hand-carved marble and stone.

In their responses to such products, critics and champions of cast iron again reflected the century's ambivalent reactions to new industrial technologies and the imitative arts. A London journal, the *Art-Union*, in its review of Coalbrookdale's display at the Manchester Exhibition of Industrial Arts, praised the company in 1846 for proving that "beauty of form may be combined with what has usually been thought the most unpromising of materials."[1] But British critics

A. W. Pugin and John Ruskin saw little that was pleasing in cast iron. As with the other manufactured wares, they denounced its inherently imitative nature and complained about the loss of honesty in craftsmanship and materials. Even worse, in their eyes, was the degrading aesthetic impact of mass-producing cheap and tawdry cast-iron versions of elegant prototypes. As Ruskin wrote in *The Seven Lamps of Architecture*, "I believe no cause to have been more active in the degeneration of our national feeling for beauty than the constant use of cast iron ornaments."[2]

The language of cast iron's critics again suggested that they had social as well as aesthetic misgivings. Their rhetoric at times reflected underlying feelings of disdain and distrust similar to nineteenth-century criticism directed at the newly prominent social arrivistes of the British mid-Victorian period and the newly rich members of America's Gilded Age, whose tastes seemed to threaten the sureties of a social and aesthetic elite.

For their own part, cast-iron manufacturers tried to ease the assimilation of this relatively new technology. Their efforts at legitimizing and dignifying cast iron often followed a pattern similar to the striving for legitimacy of social out-groups: their stance toward cast iron moved from self-effacement, assimilation, and the imitation of presumed aesthetic superiors to a blatant pride in the technology's own unique properties, resulting in a form of technological chauvinism in which manufacturers boasted that their cast-iron products were superior to the materials they had formerly emulated.

The actual history of cast iron was a long one. As a building material, it was reported to have been used in China in the sixth century B.C., and during the fourteenth century it was used for cannons and cannonballs. Fifteenth- and sixteenth-century Europe made use of cast iron for water pipes, pots, and stoves. It was in the eighteenth century, however, that developments in sand casting and changes in iron foundry procedures revolutionized the industry. In 1709 Britain's Abraham Darby first smelted iron ore with coke, making it far easier to produce cast iron on a commercial scale, and Darby's foundry at Coalbrookdale became one of the eighteenth century's most prominent producers of hollowware pots and steam engine castings.

The controversial, imitative nature of cast-iron decorative objects was, ironically, often based on technological merits: cast iron manufacturers were able to mass-produce large numbers of copies of original designs at a much lower cost than that required to produce individually created pieces. In foundries such as Coalbrookdale, company artists and draftsmen created designs for art metalworks which were then given to skilled pattern makers who carved models in wood and clay or wax when greater detail was required. Pattern makers also derived their designs from pattern books intended for wood cabinetmakers, often producing extremely intricate designs. The models were then cast in plaster from which master metal patterns were made. The metal or wooden patterns were then pressed into sand contained in two-halved rectangular frames. After the pattern was removed and the two halves joined, molten iron was then poured into the resulting cavity through channels in the sand molds.[3]

The resulting cast-iron product was then painted to prevent rusting, usually in chocolate brown or dark green, and later repainted in decorative colors or refinished to imitate bronzed or other antiqued surfaces. The Coalbrookdale catalog of 1875 advertised numerous powder and paint finishes for cast-iron hall furniture and full-size sculptures, including a "statue" finish with brown ground gold and copper, and "antique," a bronzed surface with greenish-gray ground gold, copper, and green oxide. The high-quality bronze finishes were intended for furniture used inside the house, while benches, painted in a single color, were intended for outdoor use. Also available was electrobronzing for medieval furniture: an electroplated surface finish of bronze or copper.

The love of elaborate ornament in nineteenth-century Europe and America proved to be an expensive taste, one that could be gratified much more cheaply and efficiently through cast iron. Because the original cast-iron patterns were carved in clay or wood, and because cast iron was a strong and durable material, pattern makers were able to produce extremely detailed designs based on the historicized styles then popular including neoclassical, Italian Renaissance, baroque, and rococo. Cast-iron stoves, valued for being fireproof and efficient in radiating heat, were made into decorative monumental artworks, as seen in a stately five-foot-high neoclassical stove in the shape of an urn manufactured by Britain's Carron Company in 1790, with its elegant symmetry and lofty rectangular pedestal base (fig. 45).

Cast iron first gained popularity in the 1820s and 1830s in Prussia, where manufacturers whose wares achieved a high degree of quality and finish commercialized the process for the decorative arts.[4] After sending a mine inspector to study the British iron industry, the Prussian king Frederick the Great established royal foundries at Gleiwitz

45 Carron Co., Cast-iron neoclassical stove (1790). By courtesy of the Board of Trustees of the Victoria & Albert Museum.

in 1796 and Berlin in 1804 which became the centers of Prussian decorative iron casting. The Gleiwitz foundry published its first printed catalog in 1814, which heightened the demand for decorative cast-iron statuettes, plaques, cast-iron calling cards, and even delicate necklaces. Prussian architect Karl Friedrich Schinkel, known for his spare neoclassicism in architectural design, created early elegant versions of iron chairs in 1825 with scroll back and modified cabriole legs for the gardens at the royal palaces in Berlin. By the 1830s, however, sales of cast-iron decorative objects had declined in Prussia. Zinc castings were seen as more suitable for the imitation of bronze, and by the 1870s production ceased at the Prussian foundries.

Recognizing the appeal of cast iron, British foundry owners were soon pirating Prussian designs and manufacturing their own decorative wares. For these manufacturers, cast iron was particularly appealing because of its ability to reproduce natural forms with a high degree of exactness and clarity. In an era when popular taste placed considerable value on the accurate imitation of floral and animal motifs, cast iron promised to be singularly marketable. Having begun its own manufacture of ornamental castings in 1834, including decorative iron plates, Coalbrookdale was also producing cast-iron tables with marble tops in the 1830s and cast-iron garden chairs by the 1840s. By midcentury, there was considerable demand for cast-iron domestic products in both Europe and America, encouraged by the displays of cast-iron art objects and architectural ornament at the huge Great Exhibition held in London in 1851, the New York Crystal Palace exhibition in 1853, and the Paris Exposition of 1855.

American manufacturers including J. L. Mott in New York, the New York Wire Railing Company, and Wood & Perot in Philadelphia joined British and French manufacturers in becoming prolific producers of cast-iron domestic products and furniture. The catalog of the New York Wire Railing Company in 1857 advertised architectural cornices, capitals, girders, and railings as well as garden settees, hall chairs with cabriole legs, urns, and cast-iron garden chairs decorated with morning glory motifs and grapes on vines—using patterns produced from clay models and copied from other foundries (fig. 46).[5] Popular items in both American and European catalogs were cast-iron bedsteads, which gained favor after 1850 because they were considered resistant to bedbugs and more hygienic than wooden frames.

Cast-iron manufacturers made efforts to lift their products beyond the mundane and into the prestigious realm of high art. Noted con-

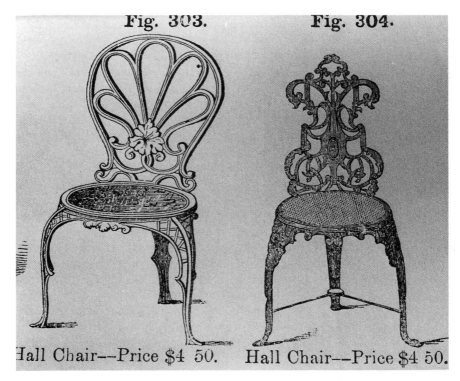

46 Cast-iron hall chairs. *A New Phase in the Iron Manufacture: Important Inventions and Improvements; Historical Sketch of Iron; Descriptive Catalogue of the Manufactures of the New York Wire Railing Company* (New York: Fowler & Wells, 1857).

temporary artists were commissioned by Coalbrookdale to produce freestanding cast-iron sculptures in imitation of bronze, including works by popular French artist Pierre-Jules Mène (1810–1879) as well as John Bell (1811–1875) of the British Royal Academy. Bell's eleven-foot marble sculpture *The Eagle Slayer*, previously exhibited at the Royal Academy, was cast by Coalbrookdale in 1848 and displayed inside Coalbrookdale's large cast-iron dome at London's Great Exhibition of 1851. Bell also provided the sculptural models for the deerhound hall table, which was shown as evidence of Coalbrookdale's artistry at the Paris Exposition of 1855. The seven-foot-long deerhound table bore a simulated marble tabletop supported by legs in the shape of life-size dogs and decorated with other emblems of the chase (fig. 47). Although some French critics were skeptical of the table's taste, Coalbrookdale continued to market cast-iron sculptures to heighten the firm's prestige.

Cast-iron manufacturers also saw a market for reproductions of museum sculptures, which could be duplicated in cast iron more cheaply than in bronze or by electrotyping. Fine-art reproductions were shown at the Great Exhibition of 1851, and by the 1870s they had achieved a high degree of technical perfection, to the extent that, as one journal reported, they even reflected "the faithful reproduction of the marks of the tool of the master hand that executed them."[6] The advantages of cast iron were seen by London's South Kensington Museum, which incorporated Renaissance and medieval reproductions manufactured by Coalbrookdale into its collections.

The nineteenth century also witnessed the increasing importance of cast iron for architectural girders, columns, and ornament, including building cornices, window arches, spandrels, and building facades. Cast iron had already been used by French architects at the end of the eighteenth century for fireproof construction in theater roofs, and by 1800 the British had developed iron frames for use in textile spinning mills, as seen in Boulton and Watt's cotton mill built at Salford in 1799, with its complete internal iron skeleton. In the

47 Coalbrookdale Co., Cast-iron deerhound hall table, designed by John Bell (1855). Ironbridge Gorge Museum Trust.

early decades of the nineteenth century, isolated cast-iron columns were used by John Nash for the Royal Pavilion at Brighton in 1818, and cast-iron skeletons were used for greenhouses and shopping galleries in the 1820s and 1830s.[7] The material provided the framework for new building types including the emergent railroad stations, which were often constructed as glass-and-iron sheds covered with masonry walls. In Paris, the Gare de l'Est contained High Renaissance cast-iron detailing, and Henri P. F. Labrouste's design for the Bibliothèque Sainte-Geneviève constructed in 1843–1850 used cast-iron interior supports covered with a masonry exterior.

Beginning in the 1850s, the varied architectural uses of cast iron became more widely recognized. Joseph Paxton dazzled the public and jolted the architectural establishment in 1851 with his cast-iron frame for the Great Exhibition's famed Crystal Palace building, which became the most famous example of a glass-and-iron structure that frankly exposed its framework without masonry cover. It was built using a prefabricated, mass-produced grid of cast-iron columns, joints, and window facades, with blue-painted stanchions and wrought-iron girders, and was covered with roof plates of glass—the largest that were manufactured at the time (fig. 48). Paxton's cast-iron design had a rapid influence, as seen in New York's Crystal Palace (1853) and the Paris Exposition building (1855). The decade also witnessed the rapid construction of cast-iron British railroad stations, including London's Paddington Station (1852–1854).

It was in the United States that cast-iron building facades received considerable commercial acceptance, becoming popular from 1850 to 1890 for stores, offices, warehouses, and hotels, many of which were concentrated in New York as well as Chicago, Baltimore, and St. Louis. In a country barely one hundred years old, and in growing industrial cities like Chicago unfettered by entrenched architectural traditions, cast-iron building facades were particularly appealing since they could be assembled quickly, were considered to be safe and economical, and provided a relatively inexpensive way to cover buildings with historicized ornamental styles that had the look of hand-carved stone. Through these cast-iron veneers, store owners lent their new establishments a look of richness and respectability.

As one of the earliest examples, engineer and manufacturer James Bogardus in 1848 covered a five-story building at 183 Broadway in Manhattan with a cast-iron facade patterned with classical Doric columns, and he became best known for manufacturing the facade of

48 Joseph Nash, *Opening the Great Exhibition: The Foreign Nave* (1851). Chromolithograph. Ironbridge Gorge Museum Trust, Elton Collection.

the Harper Brothers Building in 1854. Daniel Badger, who claimed along with Bogardus to be the first manufacturer of cast-iron facades, helped promote the practice through the publication of his catalog in 1865.

The styles of these building facades and architectural ornaments were often eclectic, including classical, northern Italian Renaissance, and baroque designs for columns, cornices, pediments, keystones, balustrades, railings, and doorway lintels. The New York Wire Railing catalog advertised several styles for ornamental column capitals, including Corinthian, Ionic, and the ornate "Corregio" style the company designed for the Saint Charles Hotel in New Orleans. Architect J. P. Gaynor's Haughwout Building on the corner of Broome Street and Broadway in Manhattan, constructed by Daniel Badger in 1857, was celebrated for its Italianate cast-iron detailing, including five stories of arched windows with Corinthian colonnettes flanked by Corinthian columns (fig. 49).

49 J. P. Gaynor, Haughwout Building (1857). Photograph by the author.

EARLY CRITICAL RESPONSE TO CAST IRON

Among design critics in Britain, the initial response in the 1840s to ornamented cast-iron designs was skepticism if not outright hostility; yet there was ambivalence, too, as critics reluctantly acknowledged the structural advantages of this emerging industrial material. The early objections were formulated in both moral and aesthetic terms, as critics complained that manufacturers of cast-iron decorative objects and architectural ornaments were being fundamentally dishonest. Chief among the early critics was A. W. Pugin, who stated his case in *The True Principles of Pointed or Christian Architecture*:

> Cast iron is a deception; it is seldom or never left as iron. It is disguised by paint, either as stone, wood, or marble. This is a mere trick, and the severity of Christian or Pointed Architecture is utterly opposed to all deception: better is it to do a little substantially and consistently with truth than to produce a great but false show.[8]

Another fundamental objection to cast iron was its repetitive nature:

> Cast iron is likewise a source of continual repetition, subversive of the variety and imagination exhibited in pointed design. A mould for casting is an expensive thing; once got, it must be worked out. Hence we see the same window in greenhouse, gate-house, church, and room; the same strawberry-leaf, sometimes perpendicular, sometimes horizontal, sometimes suspended, sometimes on end; although by the principles of pure design these various positions require to be differently treated.

Pugin's third major objection to ornamented cast iron was the tendency to overdecorate cast-iron gates, stoves, door knockers, and the like so that their fundamental purpose was too heavily concealed. Particularly objectionable was the use of miniaturized architectural detailing, ordinarily reserved for monumental scale, to ornament small, mundane utility wares. As Pugin wrote in exasperation,

> It is impossible to enumerate half the absurdities of modern metalworkers; but all these proceed from the false notion of *disguising* instead of *beautifying* articles of utility. How many objects of ordinary use are rendered monstrous and ridiculous simply because the artist, instead of seeking the *most convenient form*, and *then decorating it*, has embodied some extravagance *to conceal the real purpose for which the article has been made*!

Pugin was particularly irate at the metal-manufacturing cities of Birmingham and Sheffield, and again, his primary objection was to design dishonesty. He cited as examples door knockers and foot scrapers disguised as castles and cathedrals, "staircase turrets for inkstands, monumental crosses for light-shades, gable ends hung on handles for door-porters . . . while a pair of *pinnacles* supporting an arch is called a Gothic-pattern scraper, and a wiry compound of quatrefoils and fan tracery an abby garden-seat." Concluding his criticism, he complained that "neither relative scale, form, purpose, nor unity of style, is ever considered by those who design these abominations."

While Pugin began the early critical discourse on cast iron, it was John Ruskin who most clearly and forcefully articulated the central complaints. In his essay "The Lamp of Truth," he took a dim view of decorative cast-iron reproductions that imitated stone architectural ornament and condemned what he called "Operative Deceit" or "the substitution of cast or machine work for that of the hand," writing, "There are two reasons, both weighty, against this practice: one, that all cast and machine work is bad, as work; the other, that it is dishonest."[9] Cast iron's dishonesty lay in its clumsy pretense at being handcrafted, though Ruskin noted cast iron was always distinguishable from wrought and hammered work. The false nature of the technology, he argued, could only result in a loss of artistic integrity and a debasement of the word "precious," which could be earned only by handmade art.

Cast iron was not only tainted by its dishonesty but was also an affront to the very idea of artistic beauty. Praising medieval wrought ironwork for being "simple as it was effective, composed of leafage cut flat out of sheet iron, and twisted at the workman's will," Ruskin denounced modern cast-iron manufacture for lack of delicacy: "No ornaments, on the contrary, are so cold, clumsy, and vulgar, so essentially incapable of a fine line or shadow, as those of cast-iron."[10]

In his lecture "The Work of Iron in Nature, Art, and Policy" presented in 1858, Ruskin argued that "the utmost power of art can only be given in a material capable of receiving and retaining the influence of the subtlest touch of the human hand." He viewed iron as suitable for art only when it revealed "the action of the body and soul together," "the mingling of heart-passion with hand-power." In an era in which the accurate imitation of natural forms was a major aesthetic goal, Ruskin praised ironworkers who could duplicate the intricacies of natural form. Given his principled belief in truth to materials ("whatever material you choose to work with, your art is base if it does not bring out the distinctive qualities of that material"), he found decorative iron successful when it made use of the material's inherently "ductile and tenacious qualities"—qualities he saw as suitable for natural forms.[11] As he wrote in *The Two Paths*, "the quaint beauty and character of many natural objects such as intricate branches, grass, foliage" is "sculpturally expressible in iron only, and in iron would be majestic and impressive in the highest degree" if "rightly treated." But, he noted scornfully, "we *cast* our iron into bars—brittle, though an inch thick," and "consider fences, and other work, made of such materials decorative!"[12]

Amidst their carefully articulated objections to cast-iron ornament, Pugin and Ruskin revealed their underlying uneasiness that the new technology would undermine class distinctions. Wrote Ruskin in *The Seven Lamps of Architecture*: "yet exactly as a woman of feeling would not wear false jewels, so would a builder of honor disdain false ornaments." These ornaments, he insisted, using language laden with class associations, "are things in which we can never take just pride or pleasure, and must never be employed in any place wherein they might either themselves obtain the credit of being other and better than they are, or be associated with the thoroughly downright work to which it would be a disgrace to be found in their company."[13]

Pugin himself, when he complained about the deceptive nature of cast iron "disguised by paint, to resemble stone, wood or marble," and denounced this aesthetic deception as a "mere trick," also revealed a profound hostility to the social pretensions of the industrial material. Manufacturers who engaged in the falsehood of disguising materials, he argued, would be well advised "to do a little substantially and consistently with truth than to produce a great but false show."

If Pugin and Ruskin were scornful of cast iron's aesthetic dishonesty, journals such as Britain's *Art-Union* took a more optimistic view. In a series of articles published in 1846, the journal provided one of the earliest in-depth assessments of the developing industry. Reviewing Coalbrookdale's products on display at the Manchester Exhibition of British Industrial Art, the *Art-Union* praised a garden chair designed by Charles Crooke ornamented with intertwined flowers and leaves, as well as the firm's vases, and noted that exhibition visitors would be "surprised to learn that all these articles are made of cast iron; they receive from the moulds a degree of delicacy and refinement of which one would imagine the material incapable—being as sharp in outline and as 'neat' in all their points as they could have been if produced in the purest bronze."[14]

A sign of the *Art-Union's* sanguine view of the iron industry was its picturesque description of Coalbrookdale's ironworks. In a narrative tour of Coalbrookdale, a second article in 1846 acknowledged the obvious contrast between the glowing light of forge flames and the "tranquil softness" of the pastoral landscape. But rather than indulging in a Wordsworthian lament for a lost landscape, the journal welcomed the imperatives of industry, which turned deadly forge flames into a source of aesthetic beauty and agricultural growth:

> But this is one of the cases in which statistical science is directly at variance with poetic feeling. These flaming furnaces, instead of wasting and destroying, have been the agents which clothed the hills with beauty, and blessed the fields with fertility. They have wrenched treasures from the depths of the earth to spread loveliness on its surface, and agriculture has flourished around because manufacture was active in the centre.[15]

The *Art-Union* praised Coalbrookdale's decorative art wares for their technical perfection and their ability to compete with "the most perfect bronzes." Since cast-iron production did not allow for using chaser's tools to correct imperfections, it was the technical achievement of producing blemish-free castings that particularly impressed the journal: "An iron-casting is therefore a more wondrous work of mechanical art than a cast of brass or bronze."

In an age that considered elaborate ornamentation both beautiful and necessary, the ability of cast-iron manufacturers to ornament ordinarily plain pieces was considered a remarkable design achievement by the *Art-Union* editors, who praised Coalbrookdale for finding new applications for decorative art, including hat and umbrella stands and a florid inkstand encrusted with decorative curvilinear patterns and topped with a Grecian urn. The journal noted, "It is in these articles of ordinary use, which the great mass of mankind regard as in some way or other unsuited to ornamentation, that the taste and talent of the designer are best evidenced."[16] Summing up its praise, the journal concluded: "we cannot hesitate to express our opinion that no material is so desirable for the purpose to which it is here applied."

As to the aesthetic merit of cast-iron ornamentation, the *Art-Union* in 1846, with some exceptions, frequently used the words "artistic" and "beautiful" in its descriptions of Coalbrookdale products. Cast-iron tables were praised for their beauty and safety due to their weight, and special praise was given to a basket for calling cards decorated with a floral design: "The execution of the foliage and flowers was wondrously perfect, and we have rarely seen a more artistic ornament for the drawing room table."[17] The calling-card basket in a Victorian home signified a passageway to the ritual entry to the home interior. That the *Art-Union* gave praise to a cast-iron basket suggests a growing acceptance, a legitimization of the industrial material for home use. The cast-iron floral motifs also proved finely adapted to the Victorian love of natural forms, and the *Art-Union* found the basket's ornamentation "perfectly natural, and as it should be, imitative rather than deceptive." But the pervasive use of naturalistic motifs in cast-iron design also drew occasional fire from critics who complained about incongruous couplings of vegetation and animal imagery. As an example of "grotesque pattern," the *Art-Union* cited a Coalbrookdale garden chair with arms entwined with leaves and a coiling serpent strangely culminating in a dog's head.

While the *Art-Union* had little to say about cast-iron architectural ornament in 1846, the *Journal of Design and Manufactures* in 1850–1851 presented one of the earliest extended journalistic examinations of the quality of cast-iron architectural design. The unsigned article, written by M. Digby Wyatt, begrudgingly accepted cast iron as a building material but balked at what Wyatt viewed as abysmal taste in cast-iron architectural ornament. The article, "Iron-Work and the Principles of Its Treatment," reluctantly acknowledged the structural strengths of the material:

> Modern science having suggested iron as admirably adapted by its strength, elasticity, toughness, and durability, for the purposes of construction, . . . and our necessity imperatively urging the adoption of a material possessing such qualities, uncongenial as it may appear to our previous notions of architectural taste, we have been compelled to place it on the list of available building materials.[18]

But when it examined contemporary design, the article complained, as had Pugin earlier, of the repetitive nature of cast-iron patterns, and as with Pugin, the rhetoric pointed to a preoccupation with social overreaching:

> when we look at the monstrosities of form in every style of design, issuing from the foundries, not to be once used, and then perhaps lost

sight of in obscurity, but multiplied and perpetuated through the medium of casting—thrust forward on all occasions as if in vainglorious consciousness of their superlative violations of the laws of harmony and proportion, we cannot but feel conscious, that the steady course of utility has, at least in this instance, outstripped its more poetical associates, inventive genius and good taste.[19]

Wyatt also added a "few points of caution" to architects, echoing Pugin's injunction against imitation: "Never," he warned, "imitate in iron ornament particularly identified with stone or marble." The final item on the list of proscriptions again made clear the preoccupation with social class that accompanied aesthetic pronouncements: "Above all," Wyatt insisted, the architect using cast iron must never "make his iron-work obtrusive" or "use any ornament to repletion." Restraint was necessary, for cast-iron ornament could never mask baseness or improve the status of a lowly building: "it would be a great mistake to imagine that a so-called 'rich' cast railing" could make a poor building "look handsomer; like the beggar's cloak, it only draws attention the more readily to the poverty it almost invariably fails to conceal."[20]

Although the ostensible concern here was with the aesthetic debasement produced by endless repetition, it was the "vainglorious" misplaced pride and pretensions of cast-iron reproductions that were particularly galling—the social temerity of industrial ornament obviously mimicking its handcrafted aesthetic superiors.

In the *Journal*'s article, there was something both demeaning and diminishing about the existence of cast-iron reproductions. They seemed to debase the integrity of the hand-created original ornamentation, particularly when drawing from the traditions of classical Greek architecture. An "enthusiast for Grecian architecture" would be utterly dismayed at the sight of "a cast-iron girder thrown across an opening, a Grecian entablature and Grecian ornament stuck upon its face, and not a semblance of a Greek column in the whole composition." Mocking the popularity of neoclassical cast-iron ornament, the article denounced the resulting debasement in design in which "the graceful honeysuckle of the Erechtheion" was "shrunk in the atrophy of cast iron"—a phenomenon that was too often seen "decorating the areas and balconies of houses and buildings of every description of purpose and every gradation of ugliness."[21]

The article's denunciation of cast-iron reproductions clearly was aimed at more than poor uses of aesthetic mimicry: a building with "Grecian ornament stuck upon its face" became a woeful example of a social veneer or mask—the very embodiment of both aesthetic and social superficiality as well as pretentiousness. The social aping, it was suggested, could be nothing but ugly.

THE AMERICAN RESPONSE TO CAST IRON AFTER MIDCENTURY

After midcentury, the early critical response to cast iron presented by Pugin and Ruskin continued to shape the debate that still swirled around this material. Their complaints were echoed by nineteenth-century critics who were often reluctant to sanction commercial iron castings. But the social impact of iron castings also provoked ambivalent response as critics variously anguished over a threatened loss of design uniqueness and welcomed the uplifting influence of "tasteful" ornamentation on popular taste.

American cast-iron manufacturers, meanwhile, stepped up their efforts to legitimize new applications and were particularly enthused about being able to reproduce in cast iron the look of ornamented stone masonry. America's first cast-iron building facade appeared in 1828 on the Miner's Bank in the Pennsylvania coal-mining town of Pottsville. The architect, John Haviland of Philadelphia, reportedly had intended to use decorative stone to cover the building but, finding no local quarry, used cast-iron plates instead. The building gained wide recognition when it was illustrated in Haviland's *Improved and Enlarged Edition of Biddle's Young Carpenter's Assistant*, which, until the Civil War, was one of the most influential American handbooks on building construction.[22]

By 1848, New York cast-iron manufacturer James Bogardus had produced the city's first cast-iron facade, which covered the Milhau drugstore at 183 Broadway and was ornamented with iron Roman Doric columns at each story and a cornice frieze. In 1849, Bogardus erected additional iron fronts for the Laing stores and his own factory. The manufacturer's enthusiasm for cast-iron facades was echoed in an article in New York's *Evening Post* that outlined many of the central arguments in favor of the new facades, including their strength, quick and easy assembly, easy dismantlement, their ability to admit more light due to the strength of iron columns that minimized the need for brick walls, and their ability to allow ornamentation: "They combine beauty with strength, . . . for panels can be filled with [ornamental] figures."[23]

It was James Bogardus himself who presented some of the strongest arguments in favor of cast iron and who became one of cast iron's greatest champions. While not the first to construct cast-iron houses or storefronts (cast-iron Parisian shopfronts existed in the 1830s), he laid broad claims to having constructed "the first complete cast iron edifice ever built in America or in the world," a claim that was probably exaggerated since his factory reportedly contained some timber interior supports.[24] In spite of some exaggeration, Bogardus did help to produce a small boom in cast-iron facades not only in New York but in Philadelphia, Chicago, and other American cities as well.

Bogardus's strongest promotion of cast iron appeared in his pamphlet *Cast Iron Buildings: Their Construction and Advantages*, which, he acknowledged, was written by John W. Thompson and published in New York in 1856. Engineers such as Britain's Thomas Tredgold and others had earlier extolled cast iron's considerable strength and durability, but it was Bogardus who highlighted the material's aesthetic and moral capabilities so dear to nineteenth-century sensibilities: its ability to imitate in precise detail historicized ornamentation and natural forms in a manner not possible in stone, and its contribution to the elevation of public taste. His pamphlet reflected a growing movement by American and European manufacturers after 1850 to legitimize the new uses of this technology and engage in what might be called "technological chauvinism"—arguing that cast iron, rather than being vulgar, dishonest (as Ruskin insisted), and an unaesthetic imitation of more prestigious marble, stone, and bronze, was, in fact, superior to these traditional materials in architecture and the fine arts. Citing the material's "safety, durability, and economy," he insisted that if the public were "fully aware of its great advantages, cast-iron would be employed, for superior buildings, in every case, in preference to granite, marble, freestone, or brick."[25]

Bogardus's pamphlet mapped out the recurrent structural and aesthetic arguments in favor of the material, arguing that when multiples of the same ornament were needed, cast-iron Corinthian columns were less costly than marble and kept their original sharpness of detail far better than more traditional materials. In an era that loved elaborate decoration, cast iron allowed for more ornate sculptural details than the handcrafted originals, details that endured longer than stone and allowed for the creation of "fluted columns and Corinthian capitals, the most elaborate carvings, and the richest designs, which the architect may have dreamed of, but did not dare represent in his plans."[26] Ignoring Ruskin's complaints about the immorality and tastelessness of imitation, Bogardus saw in cast iron the potential for a heightened morality: by producing more elaborate forms for a wider public audience, cast iron was superior to stone and marble originals in its ability to help elevate public taste. Although twentieth-century modernist architects and designers would later argue that unadorned classical simplicity was the appropriate aesthetic for a technological society, Bogardus reflected nineteenth-century America's assumption that elaborate ornamentation was a vehicle for aesthetic salvation. Through the inversions of technological chauvinism, cast-iron facades that imitated stone and were often denounced as tawdry kitsch became, in Bogardus's view, morally elevated as the superior design form:

> Ornamental architecture—which, with our limited means, is apt to be tawdry, because incomplete—thus becomes practicable; and its general introduction would greatly tend to elevate the public taste for the beautiful, and to purify and gratify one of the finest qualities of the human mind.[27]

The arguments outlined by Bogardus in 1856 were repeated in the sales catalogs issued by cast-iron manufacturers for the next several decades, notably in the catalog issued in 1865 by another prominent New York iron manufacturer, Daniel Badger, owner of the Architectural Iron Works. Badger himself claimed in his catalog to have erected "the first Iron Front in the United States," but in actuality his storefront erected in 1842 on New York's Washington Street consisted of single-story cast-iron columns and lintels rather than an entire facade. Badger's catalog listed in boldface type the benefits of his cast-iron products, which included strength, lightness of structure, facility of erection, architectural beauty, and economy or cheapness. Regarding tensile and compressive strength, he suggested that cast iron's legitimacy was complete: "The established superiority of Iron in this regard now requires no argument," for no substance "has such closeness of texture, or is equally capable of resisting immense pressure." The lightness of an iron frame was again touted as beneficial in its ability to free the structure from heavy masonry walls, allowing more window space and more light to enter the structure.[28]

Arguing for cast iron's superior contribution to architectural beauty, the catalog noted that in contrast to wood or stone, "iron is capable of finer sharpness of outline, and more elaborate ornamentation and finish." Echoing Bogardus, Badger praised cast iron's economy: "the cost of highly-wrought and beautiful forms in stone or marble, executed with the chisel, is often fatal to their use," but works in iron, because of their cheapness, are "placed within the reach of those who desire to gratify their own love of art, or cultivate the public taste."[29]

In spite of the efforts by manufacturers to legitimize cast iron for architectural use, there was yet considerable resistance among architects who were skeptical of cast iron's structural stability and its use for imitating stone masonry cornices and columns. The technology, though embraced for building facades in commercial use, was never generally accepted for domestic home facades. Iron framing, however, would later become central to the design of larger buildings.

Resistance by architects was seen in the debates that took place at the American Institute of Architects in New York in 1858. *Scientific American* one year later denounced the debates as "ridiculous" and "unsound," refuting architects' fears, for example, that cast iron was combustible. Welcoming "this new and useful material," the magazine engaged in its own technological chauvinism by claiming that cast-iron ornament had become an inspiration to handcraftsmen: "the richness and variety of its moldings have stimulated the workers in stone and marble to higher efforts of their art in buildings."[30]

In his revised *History of American Manufactures* (1868), John Leander Bishop acknowledged the resistance to cast iron, noting that "the early history of this manufacture is the history of a long continued struggle against prejudices, objections and conflicting interests," yet the writer optimistically concluded that the resistance was now over, referring to "the ultimate triumph of resolute persistence and indomitable energy on the part of its founders. The innovation was resisted until Iron buildings had been thoroughly tested, and their superiority was too manifest to be denied." In the section of his history describing Daniel Badger's Architectural Iron Works, Bishop repeated Badger's contentions that cast iron saved space, had greater capacity for ornamentation, and helped improve public taste, adding that its greater availability even had hygienic benefits, as it allowed for the "decided sanitary influence of light upon the inmates."[31]

But in spite of Bishop's certainty that cast iron's manifest superiority had triumphed in architectural minds, two written books of the 1870s, one by a British engineer and one by the noted American journalist Horace Greeley, suggest that the struggle to legitimize the technology was not yet over.

The engineer Ewing Matheson, of the British engineering firm Andrew Handyside and Company, devoted his *Works in Iron* (1873) to

50 Cast-iron column capitals. Ewing Matheson, *Works in Iron*, 2d ed. (London: E. and F. N. Spon, 1877).

expanding the reader's knowledge of cast iron's capabilities, arguing that "owing to the limited knowledge which existed as to the strength and other characteristics of iron," there "has been a tendency in designing ironwork to adopt too nearly the existing conventional rules for stone or wood." Matheson added that "even now some architects seem hardly to be aware how far cast iron may be trusted."[32]

His chief argument in favor of cast iron, in essence, was a repetition of Ruskin's and cast-iron manufacturers' contention that cast iron allowed for great precision in delicate detailing or ornament: "Some designs which are too delicate to be utilized in such materials as clay, stone, and wood, are of sufficient strength if made in cast iron. Herein lies its superiority over all other materials used in architecture." Matheson's avowed purpose, however, was to reform design practices among cast-iron manufacturers who produced the "rude appearance of many of the cheap castings supplied to builders in this country," which "have justified considerable prejudice among architects." As a correction to these unappealing "rude" designs, Matheson included an engraving of a cast-iron column capital surrounded by thin curved foliage motifs (fig. 50) that would be expensive if produced in wrought iron and ill-adapted to weather outdoors, but would retain its "crispness and delicacy of outline" if produced in cast iron.[33]

Horace Greeley and the editors of *The Great Industries of the United States* (1873) devoted several chapters to championing ornamental ironwork, in part as an answer to Ruskin's disdain for cast iron, which was referred to at the beginning of an essay on architectural ironwork dismissing Ruskin as belonging to a "sentimental school of criticism." The essay mocked "a certain class of architects" who "are disposed to object to iron in architecture for sentimental reasons, mostly such as will be found in the works of Mr. John Ruskin," and mocked their belief that "it is not proper, and, indeed, that it is sinful, to imitate in one material forms used in another; that it is wrong to put iron, for instance, into forms suitable for marble or granite." These views were dismissed as lacking in "scientific method," for they paid little heed, presumably, to the advantages of cast iron that the authors then duly enumerated in nine points, noting, however, that they had no intention of attempting "any detailed exposition of the merits of iron as an architectural material," since, they insisted, "those merits are already extensively recognized among scientific men, and are rapidly gaining reputation with all classes." Nevertheless, the essay acknowledged that the process of legitimization was not yet complete, noting, "It seems singular that our architects should not have recognized that iron is a legitimate building material, but should be used legitimately, especially when the Crystal Palace has afforded so notable an instance of its value and right use."[34]

Greeley's reviews represented a paean to the new technology and exuded nineteenth-century America's continuing chauvinistic praise for cast iron's aesthetic superiority. Iron volutes were described as "more beautiful than any which can be cut from stone by the subtlest art of the sculptor," and cast iron's strength helped create "tracery of a light and graceful effect altogether beyond what is possible in wood or stone." With unabashed hyperbole, Greeley and the editors insisted that iron itself is "the chief precious metal. It can be made, even for the most delicate purposes, many fold more valuable than gold."[35]

Greeley's version of technological chauvinism included the popular nineteenth-century American argument that the new technology had the moral benefit of bringing cheap copies of expensive and beautiful art to a mass market. As if answering Ruskin's complaints about the immorality of imitating handcrafted wares and his praise for the moral purity of medieval craftsmanship, American iron foundries, particularly that of Robert Wood in Philadelphia, were highly praised: "Much of the work which now emanates from American shops compares favorably with the very best of the middle ages work in Europe in all respects, and is given to the public at cheaper rates, thus, carrying the comforts and solaces of a fine art into a large number of houses and homes." The ironwork of "American artisans" was "equal to the best of the old masters." The praise of cast iron waxed ever more poetic in efforts to prove the moral superiority of the material. If the "blessing of iron" were removed, Greeley and the editors warned ominously, "moral chaos would ensue" not only because people needed iron plows and axles for their survival but because iron, promoting the love of beauty in the form of ornament, "plays as large a part in the advancement of man's moral nature as it enacts in his physical preservation and well-being."[36]

THE QUESTION OF AUTHENTICITY

The sense of optimism reflected in the notion that cast iron would bring aesthetic and moral improvement never managed to quell the century's enduring ambivalence about the imitative nature of new technologies. Strong objections continued to be raised about cast iron's mimicry of stone and marble, and were given added impetus by the century's mounting concern for "engineering honesty" and "truth in materials." The practice of painting cast iron to resemble stone—sanctioned in structures such as J. P. Gaynor's Haughwout Building, which was painted a sandy color—was increasingly criticized as violating canons of honesty by disguising the "true" nature of cast iron.

Pugin's irritation with the practice of painting iron had stemmed as much from social disdain as aesthetic contempt, and even Ruskin, who ostensibly rejected cast iron on the basis of its aesthetic deceit, had evidenced social disdain when he dismissed cast-iron architectural ornaments as "these vulgar and cheap substitutes for real decoration." It was again America's writers, however, who often seemed relatively unconcerned with vulgarity and social pretension and who focused more directly on the issue of truth in materials. One of America's leading tastemaking magazines, *Appletons' Journal*, wrote in 1871:

> Iron, when used in architecture in imitation of stone, is an abomination. No dexterity in painting it can give it the texture or quality of

stone, and the imitation is always offensively apparent. But iron used as iron, in accordance with rules and after methods derived from its special nature, is quite another thing.[37]

That even engineers and architects were becoming self-conscious and self-critical about deception was suggested by a column titled "Bad Taste" which appeared in the journal *The Engineer, Architect and Surveyor* published in Chicago in 1874. "Too often," the article complained, "cast iron is disguised so as to appear as much like stone as possible, even to the form and amount of undercut in the mouldings, enough to deceive if possible even 'the very elect' among architects, which we hold is dishonest in principle. Honesty in design is as much to be followed as the same principle in ethics, and should be always held as one of the most necessary among all that form the principles of good taste."[38]

The painting and coating of cast iron with elaborate finishes to resemble other materials was a particularly controversial feature of the technology. In the casting process, molten metal fused with sand on the surface of the mold, creating a skin with the appearance of a silicate and a tendency to rust, necessitating a coat of paint for rust protection. The most common colors for ironwork leaving the foundry were red, brown-red, and ocher, these being the cheapest pigments. Manufacturers, however, quick to exploit iron's receptiveness to paint and elaborate finishes, covered cast-iron domestic products and architectural ornament with a full spectrum of colors, gilding, and bronze coatings. The Coalbrookdale catalog of 1850 listed among its varieties of bronzing "Antique" (greenish-gray with gold, copper, and green oxide), "Moresque," "Florentine" (rich brown with copper), as well as an "Electro" bronze creating the effect of brass or copper and "Illumination," a flat finish with unburnished gold suitable for the company's medieval-design chairs by Christopher Dresser.

Early responses to the painting of cast iron were skeptical. *Appletons' Journal* of 1869 referred to "the objectionable nature of iron as a material for architecture. Its surface requires to be painted, and paint is either glittering in crude freshness, or dingy with stains and dust."[39] Underlying *Appletons'* aesthetic views is a subtext of social commentary, for the analysis of the new technology is shaped by the perspective of status and class: paint's glittering "crude freshness" suggests the vulgarity of a social parvenu. But while British critics shunned the imitative arts for encroaching on the realm of the elite,

Appletons' views reflected a middle class anxious to protect their newly acquired status and, as *Appletons'* lofty tone suggests, put themselves far above any hint, or memory, of baseness.

In keeping with its readers' aspirations to gentility, *Appletons'* thus approached the new technology of cast iron with an air of cultivated, presumably aristocratic sensibility. Yet the journal's intent was not to dismiss the new technology but to civilize it. Having contemptuously described cast iron's glitter and dingy surface, the journal praised the possibility of the "new European technology" of coating iron with a "bronze-like surface" using melted sulfur and lampblack. This process, argued the journal, "would no doubt, give iron in architecture a much more presentable exterior than it assumes under the paint-brush." Answering critics and assuming the mantle of taste arbiter, *Appletons'* framed its response in terms of social class considerations, preferring the appearance of sedate wealth to "showy" pretension: "Possibly, however, with our national fondness for white paint, a bronze surface would not be considered showy or staring enough; but assuredly, to the cultivated eye a front of a rich though sober tint would be preferable to the soiled and rusty appearance which most of our iron buildings now present."[40]

If *Appletons'* of 1869 attempted to civilize the new technology by rendering it appropriately tasteful and dignified, Ewing Matheson's approach as an engineer in *Works in Iron* was to transform or reform the technology by applying rules of engineering common sense. Matheson was skeptical of the new electrobronzing process of coating cast iron to produce an imitation of other, more costly materials, rejecting the process not on moral or aesthetic grounds but simply because, as he noted, the process did not work well. It was better, he argued, to make castings of solid bronze or zinc. Matheson's commentary on cast iron continued to reflect the author's engineering aesthetic through his insistence on structural honesty and truth in materials: "Paint should not be used to disguise the character of ironwork, but should, as far as possible in each case, accord with its position and purpose." Rather than rejecting painted ironwork, however, Matheson argued that the engineer's aesthetic should be based on color psychology and logic: "Columns or massive ironwork, evidently sustaining a great weight, should not have an artificially light appearance given to them," but rather be painted dark green, indigo green, or dark brown. Ironwork such as ornamental capitals on columns should have lighter tints. In an era that lavished gold leaf gilding on architectural ornament to evoke an aura of wealth, Matheson's aesthetic was one of restraint: gilding "should be so arranged as to assist rather than hide the light and shade of mouldings and other parts in relief. A few gilt mouldings or a few gold lines on a spandrel or a panel, on leaves, rosettes, or rivet heads, will have a better effect and suggest greater richness and value in the gold than if it is used profusely."[41]

Contemporary writings continuously reflected an effort to apply such an engineering aesthetic to the new uses of decorative cast iron, emphasizing cast iron's integrity and legitimacy as a building material. *Appletons' Journal* of 1871 noted that ironwork being a new technology, it was "not surprising" that "at first it should have been employed simply as a substitute for stone," but praised its own New York headquarters building on Broadway between Prince and Spring streets as "the only one that distinctly obeys the law of its own material," in contrast to other buildings on Broadway with cast-iron facades constructed after stone models and painted to look like marble. Cast-iron structures that imitated the massiveness of marble or granite in an effort to lend iron a degree of architectural dignity resulted in "nothing more than so much greater surface of paint." In contrast, the Appleton Building was said to emphasize ironwork's own special nature by being "light, spacious, and graceful," painted not in imitation of marble but "in a neutral tint and picked out with gold." The journal, like Ewing Matheson, again assumed that the decorative painting of ironwork was not a camouflage but in keeping with the material's own nature: iron "admits of elaborate decoration, of tints and colors, which in stone would be atrocious." Citing New York's Crystal Palace building of 1853 "with its many tinted columns, which was so much admired," *Appletons'* made its heady prophecy for an era of technological dazzle on Broadway: "Ere many years it will be as resplendent as a fairy palace, lined with light aërial structures, gay with color and glittering with glass—a street of crystal palaces."[42]

CAST-IRON FURNITURE: A PARADIGM

The ongoing dialogue about cast iron's appropriateness for architecture and the struggle to forge an appropriate aesthetic for its use testified to the disruptions produced as new industrial materials were being assimilated into nineteenth-century frameworks of art and design. A paradigm for success in this assimilation process was the large-scale introduction of cast-iron furniture for domestic use.

Among the varieties of nineteenth-century cast-iron furniture, garden furniture proved the most enduring in popularity and the easiest to assimilate, undoubtedly because the material's solidity and durability made it suitable for outdoor use. But by 1850, manufacturers were mass-producing cast-iron furniture not only for gardens but also for middle-class home interiors. The response of critics and the public to this interiorization provided a telling commentary on the extent to which cast iron was able to achieve a degree of legitimacy.

Cast iron was early seen as particularly suitable for "cottage furniture," defined in Victorian England and America as furniture for middle-class country homes. This use of cast-iron chairs received a boost from one of the century's most prominent tastemakers, John Claudius Loudon (1783–1843), whose *Encyclopaedia of Cottage, Farm, and Villa Architecture and Furniture* published in London in 1833 became well known in both England and the United States. Loudon, a landscape gardener and horticulturist, provided a copybook with broad influence, and his taste for "Old English" designs helped popularize the commercial Gothic style. In 1833, cast-iron chairs were a rarity, but Loudon's *Encyclopaedia* contained illustrations of Gothic-style hall chairs that were praised as being "strong, durable, and cheap."[43] The Gothic chairs were to be cast in three pieces and then riveted together, and were described as hall furniture for "an inn, or even a villa," the word "even" here suggesting that cast-iron hall chairs were only beginning to be conceived of as suitable for the inner sanctum of household use.

For all their misgivings about cast iron, British and American journal critics nevertheless helped legitimize the technology by suggesting to their readers that the new castings could fit comfortably into home interiors without incongruity. Britain's *Art Journal*, in its catalog of the Great Exhibition of 1851, illustrated a London-made cast-iron stove in the stylized shape of a neoclassical urn and argued, as if in answer to skeptics, that the stove "in no degree detracts from the elegance of the most classically-furnished hall, or other apartment, in which it may be placed." The classical style itself may have played a role in helping the legitimization process, for it endowed this newly arrived technology with a degree of aesthetic gentility and timelessness: as the *Journal* noted, the "antique urn" was "another proof of the great and universal applicability of the graceful designs of antiquity," forms that "are capable of being reproduced for new purposes, unthought of by the men who imagined them, but whose pure taste has rendered their ideas immortal."[44]

Critics were similarly sanguine about cast-iron "art manufactures" and contributed to the technology's assimilation even into the intimate recesses of the home. In its review of the Paris International Exhibition of 1879, the British and American editions of the *Art Journal* praised a Parisian cast-iron fountain in terms of the material's intrinsic properties, noting that the designs were "sharp and brilliant." In a conscious effort to sanctify the cast iron, the *Journal* proclaimed that such fountains were "in truth, examples of high Art, usually adornments for grounds and gardens, but such as might grace halls, drawing rooms, and boudoirs."[45] That cast iron was deemed suitable even for boudoirs was a suggestion to the Victorian reader that in an era of heavily upholstered furniture, objects such as the fountain had a suitable degree of delicacy and femininity for a room associated with a woman's sensibility.

British and American buyers who did indeed incorporate cast-iron furniture into their sitting rooms, hallways, and boudoirs provided not only a reflection of the century's enthusiasm for household products representing industrial progress but also a reminder that the new cast-iron type forms such as highly-ornamented hallway furniture coincided with Victorian tastes and social values. Furnishings specifically designated for hallways became particularly popular in Victorian England and America near the middle of the nineteenth century and declined by the beginning of the twentieth. The Coalbrookdale Iron Company catalog of 1860 and the New York Wire Rail Company catalog of 1857 included a variety of cast-iron items designated for use in the home hallway such as hat racks, umbrella stands, hall tables, hall chairs, and an important Victorian innovation—the hall stand, which combined several of these utilitarian functions (fig. 51).

Cast-iron furniture, however, appears to have served more than utilitarian functions by helping to fulfill the cultural meanings associated with the hallway itself. Historians of Victorian England and America have noted the symbolic and ceremonial meanings of the front hallway, a place intended to convey an aura of status, stability, and dignity.[46] In his editor's notes to the 1876 American edition of Charles Eastlake's highly influential *Hints on Household Taste*, the American editor Charles Perkins noted that in houses the hallway "takes the place of a Roman *atrium* or the Moorish 'pateo,'" becoming a place well suited for trophies, arms, artworks, as well as "a few chairs of ample proportions." He concluded, "Nothing adds so much to the noble appearance of a house as a wide and lofty hall."[47]

51 Coalbrookdale Co., Cast-iron hall stand. Coalbrookdale Company Catalogue, Section II, 1875. Ironbridge Gorge Museum Trust.

The hallway had other social functions laden with status meanings, being a place that conducted social peers of the middle-class or upper-middle-class home owner toward more formal space in the home, yet also a place where social inferiors remained until directed elsewhere so as not to intrude on the family. Thus hallway furniture, which was made of wood as well as cast iron, was often designed with an attempt at grandeur intended to reflect the high social standing of the household. The hall stand, which gained prominence in the United States in the 1870s and appeared somewhat earlier in Britain, combined hat rack, umbrella stand, marble-covered table, and looking glass, and since it was not often found in lower-class homes, was itself a mark of social standing. Umbrellas themselves were, for Victorians, a bourgeois emblem of respectability, and mirrors made of plate glass, which was still expensive in the nineteenth century, were considered a sign of wealth and status. Marble also, though its use had become somewhat debased through widespread availability, still had cultural associations with luxury.

Cast-iron hall chairs, with their ornateness and heavy stability, however uncomfortable they may have been to sit on, admirably served the utilitarian needs and social functions of a Victorian home. Hall chairs were often not meant for comfort but served instead as emblems for social status: people kept waiting were social inferiors and hence their comfort was not an issue, while peers were shown into the home's formal rooms. Hall furniture also had the appearance of solidity. Since the hallway was often simply a conduit to other rooms, it was not considered necessary to make the furniture light and easily movable. Contemporary writers noted also that hall chairs were designed to evoke dignity and an atmosphere of tradition and stasis. Charles Eastlake in his revised *Hints on Household Taste* (1869) wrote that entrance hall furniture was expected to give "an appearance of solidity" in order to convey "respect for early traditions of art," unlike the dining room, which "may have succumbed to the influence of fashion in its upholstery," and the drawing room, which "may be crowded with silly knick-knacks, crazy chairs and tables, and all those shapeless extravagances which pass for elegance in the nineteenth century."[48]

Manufacturers themselves made a deliberate and conscious effort to lend their cast-iron furniture a degree of dignity, tradition, and status by employing well-known artists as designers who heightened the prestige of the firm and cast iron itself. The Coalbrookdale Company employed a number of well-known contemporary artists in ad-

52 Coalbrookdale Co., Cast-iron garden bench (c. 1875). Water plant design by Christopher Dresser. Ironbridge Gorge Museum Trust.

dition to its staff designers to produce patterns for furniture and the ornamental bas-relief panels that adorned the backs of cast-iron chairs. By 1869, the firm was manufacturing hall and garden chairs designed by Christopher Dresser, who had already gained recognition in England as a lecturer and professor of botany as well as the author of *The Art of Decorative Design* (1862) and *The Principles of Decorative Design* (1863). Coalbrookdale included in its catalog of 1871 an elaborately designed medieval high-back cast-iron chair with electrobronzed finish, designating the designer as Dresser with decorative roundel by J. Moyr-Smith.

Dresser's design aesthetic emphasized flat, abstracted patterning rather than the sculptural treatment of plant forms that often characterized Coalbrookdale's cast-iron furniture. The characteristic simplicity and elegance of his geometric stylizations could be seen in his garden and hall furniture, including the back panel for his Coalbrookdale "Water Plant" bench of 1875 and his cast-iron chair of 1875 with stylized Gothic decoration of foliation and spiky leaves (fig. 52).

Dresser's views on cast-iron design were articulated in his report on London's International Exhibition of 1862, in which he praised Coalbrookdale's manufactures, citing the firm's castings of leaves as "excellent studies for the treatment of plants in ironwork not that they are to be applied imitatively" but rather treated in a "conventional manner."[49] His emphasis on stylization appeared the same year in *The Art of Decorative Design*, in which he argued, "Conventionalized plants, we say, will be found to be nature delineated in her purest or typical form, hence they are not imitations, but are the embodiment in form of a mental idea of the perfect plant."[50]

Dresser's aesthetic, his call for functional utility and "conventionalization," was based in part on his own observations of plant growth. What he saw as a botanist were admirable examples of the principles of adaptation and utility, principles that he described in a series of articles on botany published in the *Art Journal* in 1857. But it was the lectures and writings of architect Owen Jones, especially Jones's *Grammar of Ornament* (1856), that were the most influential, particularly his proposition 13, which stated that "flowers or other natural objects should not be used as ornament, but conventional representations founded upon them sufficiently suggestive to convey the intended image to the mind, without destroying the unity of the object they are employed to decorate."[51] Jones's concluding chapter on "leaves and flowers from nature" included a color plate by Dresser illustrating the geometric arrangement of flowers—a design that closely resembled the flat, abstracted flower patterns Dresser later created for Coalbrookdale.

As an industrial designer, Dresser countered the Victorian assumption that equated highly elaborate ornamentation with an expensive appearance, and he argued for a truer elegance based on simplicity. In his 1862 exhibition report on ironwork, he praised a Sheffield iron stove fender for being "costly, yet neat; and it does not look extravagant, although the surface is engraved"—the word "neat," in contemporary usage, connoting simplicity.[52] Unlike William Morris, whose design reforms of the 1880s entailed costly labor and materials, Dresser, through the use of cast iron and electroplating, helped make his designs available to a larger market. As he wrote in *Principles of Decorative Design* (1873), "if the designer forms works which are expensive, he places them beyond the reach of those who might otherwise enjoy them."[53] Dresser's "Water Plant" garden seat was still available in Coalbrookdale's catalog of 1890 and sold for sixty-eight pounds, the same price as his medieval high-back chair. By Coalbrookdale's standards, Dresser's chairs were not expensive in comparison to the firm's other, more elaborate designs, which cost one hundred pounds each.

While never fully satisfying critics' objections to their imitative nature, cast-iron ornamental and domestic products continued through the end of the century to receive praise for their most beautiful manifestations in the hands of masterful designers such as Dresser and the American architect Louis Sullivan, whose cast-iron art nouveau architectural ornament lent grace and elegance to the masonry facade of Chicago's Carson Pirie Scott store (1899–1901) and the interior staircase of the Chicago Stock Exchange building (1893). By then, decorative cast iron had found its place not only on building exteriors and as outside door knockers and foot scrapers but also within the home as decorative vases, plates, and hall furniture.

That cast iron had achieved a degree of legitimacy suggested that it fulfilled a number of important social functions. By serving as an exterior facade for masonry buildings and as interior hall furniture for ceremonially greeting guests to the home, cast iron helped meet the needs of manufacturers and middle-class consumers for an impressive social image, a dignified front. The wide array of historical styles readily available through cast-iron reproductions—the neoclassical cast-iron vases, medieval and rococo cast-iron chairs—helped provide a connection to the past in an era of social flux and gave further status to a newly arrived technology as well as to consumers with social aspirations. Even the heavy weight of cast iron may have provided a source of security and stability in a century when so many aesthetic and social frameworks were under siege. Ultimately, the assimilation of cast iron in spite of—or perhaps because of—its pretensions and imitative nature testified to the century's ongoing need to lend glamour and stature to its technological achievements, and to the public's optimistic eagerness to embrace these new technologies in the name of social and aesthetic progress.

Chapter 6

Classicizing the Machine
Ornamented Steam Engine Frames and the Search for an Industrial Style

In her poem "Coalbrook Dale" written in 1785, British writer Anna Seward (1747–1809) mourned the despoliation of the pastoral landscape in Shropshire where the ironworks at the town of Coalbrookdale had violated the rural countryside with "rattling forges" and "hammer's din." Yet while evoking the image of an industrial assault, Seward in her poem also wrote admiringly of Birmingham, a city with "rich inventive commerce" and science that "with great design" was developing the steam engine. Seward's paean to the steam engine echoed the era's infatuation with technological giganticism as she envisioned the

> . . . vast engine, whose extended arms,
> Heavy and huge, on the soft-seeming breath
> Of the hot steam, rise slowly.[1]

Just as Erasmus Darwin, her friend and contemporary, had delighted in the steam engine's "large limbs" and "giant-birth," Seward stood in awe of this "vast engine"; yet it was this same gigantic stature that would seem fearsome in the eyes of later observers who pondered ways to master the massive machine.

During the next century, the steam engine increasingly came to symbolize the nineteenth century's great pride in engineering achievement and its concurrent anxieties about steam power. James Watt's technical advances in engine design—his patents for a separate steam condenser in 1769 and rotative steam engine in 1782—helped generate optimistic hopes for industrial expansion, yet his development of a governor to control engine speed was a mechanical as well as metaphoric reminder that new technologies required constant vigilance and containment to prevent their explosion or escape from control.

For early nineteenth-century manufacturers, pattern makers, and mechanical engineers who were engaged in designing steam engines, the major task was to solve technical problems, but they also considered the outer appearance of the engine. The design history of stationary steam engine frames could be seen as emblematic of the search by nineteenth-century manufacturers and engineers for a style befitting a new technological era. After 1830, when many of the early mechanical issues in engine design had been resolved, designers confronted the need to create engine frames that embodied the century's pride in technology while also quelling residual fears of the machine's potential for explosion.

Nineteenth-century designers were, in effect, creating engine frames that served a multitude of mechanical, cultural, and aesthetic functions: the frames not only served as a base to which were attached the engine's components but also, when decorated, helped mirror and magnify the machine's cultural image. Impressive architecturalized, ornamented engine framing represented not only a way to heighten the machine's status but also a way to stabilize, literally and symbolically, the disruptive tremors of technology.

The aesthetic issues of steam engine design arose only after a rapid series of pivotal developments in steam engine technology in the late eighteenth century, following Thomas Newcomen's development of an atmospheric engine for pumping water from mines in 1712. Due to its low thermal efficiency, the power-generating costs of Newcomen's engine had proved prohibitive. It was British instrument maker and inventor James Watt (1736–1819) whose patent for a separate engine condenser in 1769 helped revolutionize steam engine efficiency, paving the way for the production of engines to drive machinery in mills and factories and blowing engines used in blast furnaces for iron production. Watt joined in partnership with British manufacturer Matthew Boulton in 1775; by 1800, Boulton and Watt with their licensees had produced nearly five hundred engines.[2]

Watt's engines employed low-pressure steam. He remained skeptical of the high-pressure engines and, wary of their perceived danger, scuttled the experiments of his foreman William Murdock on a

53 Boulton and Watt rotative beam engine, "Lap Engine" (1788). Trustees of the Science Museum (London).

high-pressure locomotive engine. (It was Richard Trevithick who in 1801 produced the first successful steam road engine and in 1804 the first steam locomotive.) Watt was well aware of the need to regulate the power of steam, and indeed it was he who in 1788 first used a pendulum or "flyball" governor to regulate engine speed, an adaptation of the governors already used in windmills. Watt's automatic governor consisted of two heavy rotating weights that flew outward under centrifugal force and slightly closed the steam valve when the engine speed rose above a set limit, restoring the engine and driven machinery to a safer speed. In later steam engines, the governor's central weight sometimes received honored status by being designed as a small classical urn made of polished iron or brass, or decorated with Egyptian motifs.

Late eighteenth-century steam engines were designed with mechanical function and utility as primary considerations, and the cabinetmakers who produced the first engine frames used heavy timbers with no architectural detailing or ornamentation (fig. 53). By 1798, James Watt had begun using cast-iron columns and beams, and by

1800 cast iron was used for all the major structural components of the engine, including the walking beam, connecting rod, and the columns supporting the beam.[3] By 1820, the conventional design for rotational steam engines had been established, though engines continued to be individually designed and manufactured. The design of engines was passing from the hands of the cabinetmaker to the mechanical engineer and pattern maker, who produced the patterns in wood from which all cast-iron elements were made.

Printed treatises on the aesthetics of steam engine design were rare during the early decades of the nineteenth century, and rare even afterward. The closest to a very early formulation of an engineering aesthetic could be seen in the writings of Thomas Tredgold (1788–1829), a British engineer whose book *The Steam Engine* (1827) was widely read by engineers in England and America and was reprinted in two volumes in 1838. Tredgold became known for his efforts to establish scientific principles for engineering design and construction. Basing his writings on eighteenth-century French engineering theory, he discussed in his book the structure and proportions of engine parts, and he originated two units of measure that were to become standard in describing engine performance, including the unit for heat quantity later known as the British thermal unit.

Tredgold was also particularly interested in testing the strength of materials and structural forms used in steam engine construction. After working as a journeyman carpenter, he came to London in 1813 and spent ten years as an architect, later establishing his own architectural firm. The results of his testing of wood and cast iron as construction materials were described in two influential works: *Elementary Principles of Carpentry* (1820) and *A Practical Essay on the Strength of Cast Iron and Other Metals* (1822). His designs for cast-iron steam engines were considered a considerable improvement over the wood-frame designs of James Watt.

Tredgold's engineering aesthetic was formulated very briefly in *The Steam Engine*, where he argued for "good proportions and fine craftsmanship." Although he advocated a well-designed and even "beautiful machine," his ultimate purpose appeared to be practical and utilitarian, for as he wrote: "appropriate forms, good proportions, and excellent workmanship should be attended to in all machinery; and in many instances it is desirable that they should be beautiful, for a beautiful machine will be so attended to as to produce economy where an inferior one would perish by neglect."[4]

Tredgold's treatise of 1827 never specified the nature of "appropriate forms" and "beautiful machinery," but the engravings of steam engines included in his book suggest that his aesthetic referred to unadorned frames that had, as he wrote, "good proportions." But while the earliest eighteenth-century wooden beam engines were unornamented, in some of the early nineteenth-century cast-iron steam engines there was evidence that designers and manufacturers were conscious of aesthetic appearance and were going beyond functional requirements to include architectural and ornamental detailing.

The firm of Boulton and Watt after 1800 established a precedent for adding historicized architectural ornament to steam engines through the use of cast-iron columns in the austere classical Tuscan style. As early as 1807, the company had designed a more overtly ornamented fifty-two-horsepower beam engine with fluted Doric columns and triglyphs on the frieze for the cotton-spinning factories of James Kennedy of Manchester.[5]

Manufacturers began architecturalizing beam engines more widely after 1830, turning, as Boulton and Watt did, from Tuscan simplicity to the use of fluted classical columns most often in the Doric style. Some designers went further, producing engines that were, in essence, miniature Greek temples complete with fluted columns, entablature, triglyphs, and base. Thomas Tredgold's enlarged edition of *The Steam Engine* (1838) was published with a separate volume of illustrations which included several classicized engines, such as the stately neoclassical engine for the steamship *Tiger* with two tiers of fluted Doric columns bedecked with French Empire–style neoclassical laurel wreaths on the entablature (fig. 54).[6]

Viewing similar examples of classicized steam engines, twentieth-century design critics committed to a functionalist aesthetic were apt to ridicule the nineteenth-century practice of ornamenting engine frames. Lewis Mumford in *Technics and Civilization* (1934) denounced engineers who "in the act of recklessly deflowering the environment" had "sought to expiate their failure by adding a few sprigs or posies to the new engines they were creating: they embellished their steam engines with Doric columns or partly concealed them behind Gothic tracery." The most damning word in Mumford's indictment was "concealed," for indeed ornament to a machine age functionalist represented nothing more than a misguided effort to disguise the beauty of machines. To the functionalist, as Mumford noted, "Ornament, conceived apart from function, was as barbarous as the tattooing of the human body: the naked object, whatever it was, had its own

54 Engine for the steamship *Tiger*, Edward Bury, engineer. Thomas Tredgold, *The Steam Engine*, vol. 2 (London, 1838).

beauty."[7] It was this same hostility to concealment that prompted Herbert Read's denunciation of a British Museum model of a classicized factory beam engine dating from 1830. In *Art and Industry* (1953), Read ridiculed the engine's "absurd little cast-iron scrolls above the cornice," which "make a strange disharmony with the purely functional parts of the engine."[8]

Although twentieth-century critics may have considered the steam engines' classical ornament absurd, during the nineteenth century the decorative frames were associated with a rich overlay of cultural meanings. There were indeed numerous factors accounting for the appearance of ornamented, particularly classicized, engines. There already existed a long tradition of using ornamental, classical detailing for clocks since the Renaissance, and a precedent for ornamented and architectural frames was seen in the works of eighteenth-century instrument makers in London who produced elaborately decorated microscopes and ornamental lathes for wealthy aristocratic patrons. These craftsmen, who were often clockmakers, were accustomed to using ornamentation and familiar with classical proportions.[9]

The practice of adding an ornamental housing to scientific instruments was seen, for example, in the richly carved miniature temple facade sharing the stand with a late eighteenth-century air pump designed by John Prince, a Protestant minister, in Salem, Massachusetts, in 1783 (fig. 55). Air pumps consisted of a piston, cylinder, and glass tubes used to produce a partial vacuum and provide a scientific demonstration of, among other things, the effect of the lack of oxygen on animals or fire. Encased in a wooden cabinet ornamented with linear pediment and classical columns carved in relief, the ornamented air pump was endowed with honorific status. Its mechanism and scientific technology itself became glorified when associated with Doric columns and an entablature carved with classical triglyphs. The air pump's use of a piston and cylinder to create a vacuum would be seen in later steam engine designs, and the pump's allusions to a small classical temple hinted at the architecturalized Doric steam engine frames that became popular during the next century.

Nineteenth-century machine-tool makers who had formerly produced clocks continued the tradition of using ornamented casings with architectural details even when making innovations in tool design, as seen in Henry Maudslay's tools for mass-producing ships' blocks designed by Marc Isambard Brunel for use at the Royal Dockyard at Portsmouth.[10]

The ornamentation of nineteenth-century steam engine frames was thus not an anomaly but, in part, a continuation of a technological tradition. Even more, the frames reflected the pervasive nineteenth-century design aesthetic, which assumed that ornament was not only desirable but necessary. While Lewis Mumford championed "naked" machines, nineteenth-century steam engine designers using ornamentation often appeared to be caught up in the century's conviction that there was something barren about unclothed forms, something indecent and undressed about raw new machines. In his novel *Hard Times* (1854), which presented an incisive view of industrialism's social impact, Charles Dickens described the house of Thomas Gradgrind in disdainful terms with its "Iron clamps and girders"—a house with no ornamentation to cover its brazen mechanics: "Not the least disguise toned down or shaded off that uncompromising fact in the landscape."[11]

55 John Prince, Double-barreled air pump (detail) in fine wood case (1783). Smithsonian Institution, Washington, D.C.

Joseph Paxton, designer of the Great Exhibition's innovative Crystal Palace in London, with its glass panels and prefabricated iron frame, assumed that his unadorned structures required ornamentation. In his lecture "The Industrial Palace in Hyde Park" presented to Britain's Society of the Arts and reprinted in 1850 in *The Engineer and Machinist,* Paxton described his new design for a private house to be covered wholly in glass, but reassured his audience that "structures of this kind are also susceptible of the highest kind of ornamentation in stained glass and general painting."[12]

Although John Kouwenhoven and Herwin Schaefer have provided considerable evidence that a nineteenth-century tradition of functional or "vernacular" design coexisted with a "cultivated tradition" of decorated forms, particularly in the United States, the evidence of nineteenth-century design journals suggests that designers most often assumed the need for some degree of ornamentation.[13] Benjamin Silliman in his catalog of New York's Crystal Palace industrial exhibits of 1853 wrote of the "demand for decoration that had arisen in every branch of manufactures" and argued, "we are no longer contented with the plainness that was once satisfactory."[14] At midcentury, it was the choice of ornamental styles that was most open to debate among warring factions, often divided between proponents of Gothic and neoclassical design.

Although there were occasional examples of Gothic and even Egyptian design, neoclassicism became the prevalent style chosen by steam engine designers, who used classical detailing for a variety of engine types.[15] Architecturalized elements could be seen, for example, in "house-built" simple beam engines, which were virtually part of a building and were constructed of cast iron, which readily lent itself to decorative patterns. In Oliver Byrne's treatise *The American Engineer, Draftsman and Machinist's Assistant* (1853), a high-pressure single-column beam engine built by the J. T. Sutton Company for the Franklin Ironworks in Philadelphia, circa 1845, supported its massive beam and entablature on a fluted cast-iron column with an elegant Corinthian capital with stylized, delicate scrolls (fig. 56).[16]

Architecturalized elements were often seen in engines built for woolen and cotton mills as well as ironworks, and for blowing engines designed to compress air needed for smelting iron in blast furnaces. The Priorslee Ironworks at Lilleshall in England was equipped with a double engine called "David and Sampson" built in 1851. The installation had two cylinders, two beams, and two air vessels, and had a single crankshaft and flywheel in common. The valve gear was supported by elements of a classical Doric temple, including fluted Doric columns, entablatures with triglyphs, and pediments. Doric columns continued to be a central architectural element in single-column beam engines of the 1860s and 1870s as well as engines with four and six fluted columns. A British high-pressure compound beam en-

Classicizing the Machine 187

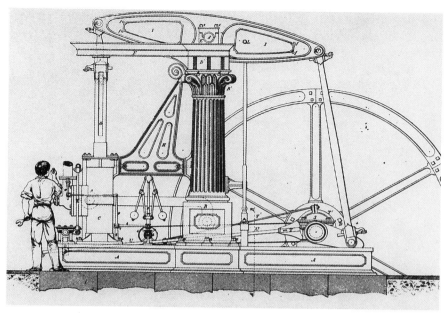

56 High-pressure steam engine constructed by J. T. Sutton and Co. for Franklin Ironworks, Philadelphia (c. 1845). Oliver Byrne, *The American Engineer, Draftsman and Machinist's Assistant* (Philadelphia, 1853).

gine designed in 1860 offered increased power and economy of action, and its innovative design was dignified by a single stately Doric column (fig. 57). The American high-pressure steam engine *Yankee Girl*, built by New York's Washington Iron Works in 1861 for use in a sugarcane mill, made a similarly imposing appearance as a Doric temple with four pedestalized columns and a long cast-iron entablature ornamented with triglyphs (fig. 58).

The choice of neoclassicism for steam engine frames was spurred by the style's popularity in architectural and home furnishing designs starting in the late eighteenth century and continuing into the nineteenth century (although a fanciful antecedent that joined classicism and steam technology could be seen as early as the work of Hero of Alexandria, whose treatise *Spiritalia seu Pneumatica* [60 A.D.] described an ingenious technique for opening and shutting temple doors using the power of heated air generated by an altar fire).[17]

The eighteenth-century interest in ancient classical culture had been reawakened by excavations at Herculaneum (starting in 1738) and at Pompeii (started in 1748) and by discoveries at Paestum, the site in southern Italy of early Greek Doric temples. Inspiring a reappraisal of classical art, Johann Joachim Winckelmann in *Thoughts on the Imitation of Greek Works* (1755) championed the classical aesthetic for its "noble simplicity and calm grandeur."

In Great Britain, a renewed interest in classical architecture was

57 Compound beam engine (1860). Trustees of the Science Museum (London).

Classicizing the Machine 189

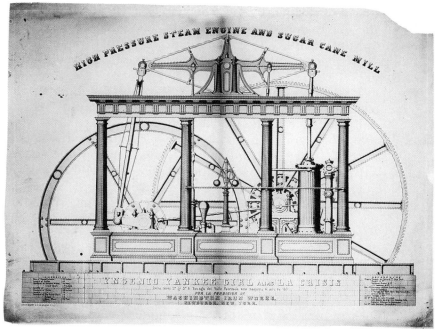

58 *Yankee Girl*, high-pressure steam engine and sugarcane mill (1861). Lithograph by Endicott & Co. (N.Y.) for Washington Iron Works. Smithsonian Institution, Washington, D.C.

seen in the architectural and interior designs of Robert and James Adam and in the work of British architect James Stuart, who helped popularize neoclassicism with the publication of *Antiquities of Athens* (two volumes in 1762 and 1789) and the building of his Doric-style Hagley Temple (1758). Neoclassicism continued as a vital architectural idiom in the nineteenth-century architecture of Britain's Sir John Soane and John Nash as well as the buildings of America's Minard Lafever and Robert Mills (best known for his Greek revival government buildings in Washington, D.C.). In 1850, American architect and tastemaker Andrew Jackson Downing, in his widely read *Architecture of Country Houses*, described the classical style as the most popular for homes.[18]

England and America also experienced a renewed interest in ancient Roman architecture and the work of Andrea Palladio, the sixteenth-century Italian architect whose Villa Rotonda and Villa Foscari had porticoes modeled after the temple fronts of ancient

Rome. The neo-Palladianism launched by Lord Burlington in England in the 1720s gained the attention of Thomas Jefferson, who, after journeying to Rome, designed Monticello (1770–1806) and pavilions at the University of Virginia adapting stylistic elements of ancient Rome.[19]

The neoclassicism of the French Empire and British Regency also became favorite styles in furniture design. In England, Thomas Hope heightened the fashionability of the Greek revival style through the publication of his book *Household Furniture and Interior Decoration* (1807), which contained engravings of his own classically inspired furniture designs based on his observations and collections of antiquities. The book also helped popularize the Greek revival style in America and was extensively quoted in Thomas Webster's *Encyclopedia of Domestic Economy*, published in New York in 1845. Hope's neoclassical bias toward the "permanent and unfading" and "repose of surface"—two fundamental attributes of classical style—was widely embraced by British furniture manufacturers through the 1830s and by Americans who continued manufacturing neoclassical furniture through the 1840s and after.[20]

The choice of neoclassicism for steam engine frames also reflected styles made popular in nineteenth-century cabinetmakers' pattern books. Often trained as cabinetmakers, steam engine designers would have had access to the most popular furniture and cast-iron pattern books of the era, including one of the earliest pattern books in America, John Hall's *Cabinet Maker's Assistant* (1840), and Lewis Cottingham's illustrated design book, the *Smith and Founder's Director* (1823), which helped popularize the classical style.[21]

But the choice of neoclassicism as the central style for architecturalized beam engine frames was not only due to contemporary notions of correct aesthetic taste; it also stemmed from a complex array of important engineering, cultural, and psychological considerations. By the 1840s, factory and mill owners had come to view their steam engines as an important ingredient in lending status and legitimacy to their establishment. Visitors were often invited to admire the engines, which were kept clean and proudly displayed as an emblem of the firm's technological modernity. The ornamented machines, sometimes set in elaborately decorated engine rooms, became, in essence, status symbols designed to look impressive. Neoclassicism (as well as the occasional Gothic and Egyptian designs) became an important means of providing the engines, as well as the firm, with an air of dignity and grandeur.

The use of neoclassicism as a design idiom for early engines was also connected to salient cultural meanings associated with the style itself. Neoclassicism contained cultural associations that were clearly separate from the functional requirements of machine design—associations of status, legitimacy, prestige, and permanence—and thus was particularly well suited to heighten the status of a new technological archetype. The choice of classical style to add a sense of social legitimacy and stature had ample political and cultural precedence. The Federal neoclassical style helped the emerging American republic establish its stability and stature at the end of the eighteenth century, and as Siegfried Giedion observed, the self-made emperor Napoleon embraced the neoclassical French Empire style to legitimize his own status. (And again in the twentieth century, a severe architectural version of neoclassicism would be chosen by the emergent Third Reich in Germany to establish its legitimacy.)[22] More directly, neoclassicism had long been associated with wealth and prestige, particularly in late eighteenth-century and early nineteenth-century architecture and furniture design.

There was also precedence for choosing classicism as a style for new technological archetypes. The eighteenth-century French architect Claude-Nicolas Ledoux drew on the imagery of neoclassicism for his design of a saltworks at Arc-et-Senans near the Forest of Chaux in 1775. He found nothing incongruous about attaching a grand entrance portico in classical Tuscan style to link a row of factory buildings, warehouses, and workshops all placed in a large semicircle.

Ledoux's classical portico signified that the visitor was about to enter the awesome world of eighteenth-century industry. Ledoux, known for his stripped classicism and austere geometries, chose the grand neoclassical portico to help establish a version of neoclassical metonymy, using one architectural element to signify the whole. A single element of neoclassical detailing could evoke the full range of cultural and architectural associations. This neoclassical metonymy would later be used by nineteenth-century designers for their own new technological archetypes.

In their designs for the earliest railroad stations in England and America, architects again turned to classical motifs to lend dignity and portent to the new technology. The earliest American railroad station was the "car house" built for the Boston and Lowell Railroad in Lowell, Massachusetts, in 1835 and designed as a small Greek temple, complete with four classical columns under a pediment with

192   *Breaking Frame*

59 Thomas Bury, *Exterior of Euston Station* (n.d.). Engraving. Trustees of the Science Museum (London).

ten columns down the side. A single track ran behind the colonnade, and the remaining space was reserved for a platform and offices. These classical references were also seen in London's famed Euston Station, built in 1835–1839, with its look of a monumental Doric temple (fig. 59) and later in the neoclassical grandeur of McKim, Mead and White's Pennsylvania Railroad Station built in New York (1906–1910).

After 1840, architects increasingly turned to more eclectic styles, but neoclassicism had clearly served important psychological functions in the early designs. By clothing the new technological archetype in traditional historicized forms, designers and architects were able to ease the introduction of the new steam-driven railroads and help familiarize them, while also establishing the prestige and legitimacy of the new means of transportation itself. The railroad station with its neoclassical temple facade became an island of security and stability—a reassuring place of departure for those about to embark on a potentially explosive new experience. Neoclassicism, resonating with associations of stability and order, helped counter fears about railroad accidents and speed.[23]

The neoclassical frames of nineteenth-century stationary steam engines gave a similar aura of stability and stasis to other steam technologies, which were viewed by the public with both awe and alarm. Amidst all the discourse about progress in early nineteenth-century newspaper stories lauding the new steam inventions and experiments, recurring articles about explosions and accidents generated anxieties about technological dangers that could be eased by echoes of classical calm. Classical frames belied the potential for faulty steam engine governors, uncontrolled speed, and exploding flywheels.

There were also other reasons for adopting the classical style. Nineteenth-century engineers were primarily faced with the task of assuring a machine's stability, with providing a heavy, secure foundation for engine parts, and in the case of mill engines with creating a framework to contain intense vibrations. Yet engineers were preoccupied with creating not only a solid and stable framework for their stationary steam engines but also a reassuring appearance of stability, which made classical design motifs all the more appropriate because of their associations with permanence, and stillness. Oliver Byrne, commenting on the Sutton single-column beam engine, wrote that the fluted column gave the engine a "compact and solid appearance" and provided a framework "free from all the jarring or tremor, so common to beam engines."[24] The engine, cloaked as neoclassical temple, suggested that the tremors of new technologies were being kept safely under control.

In a rare discussion of classical style and steam engine design, British engineer Samuel Clegg (1814–1856) wrote what was perhaps the first focused study on the aesthetics of steam engine design, a study that provided a telling commentary on early nineteenth-century efforts to reconcile engineering requirements with aesthetic concerns. Clegg's *The Architecture of Machinery*, published in London (1842), was remarkable because it addressed a new area—the aesthetics of engine design—and did so with lucidity, precision, and an engineer's expertise. The son of the famed British gas-lighting engineer Samuel Clegg (1781–1861), the younger Clegg was himself a civil engineer and professor of engineering. His *Architecture of Machinery* reflected an eye for design and a reformer's zeal, for his avowed purpose in writing the book was to improve both the engineering design and the aesthetic appearance of steam engines.[25] As he noted in the book's opening chapter, there were "numberless contortions that have crept into the 'patterns' of even some of our best and most scientific engineers." His purpose, he wrote, was twofold: to elucidate "correct principles of 'taste'" and to correct "irregularities in construction and form."[26]

Clegg's specific intent was to improve the looks of contemporary steam engines that used architecturalized cast-iron frames modeled after the patterns of classical antiquity. Clegg, who was a great admirer of classical forms, never questioned the design convention of using classicized engine frames, but he did object to examples of "deformity and burlesque"—engines with misapplied or misappropriated classical design imagery, and machine designs that violated classical rules of correct proportion.

As an example, he cited engines with a version of Empire-style ornamentation that were "elaborately panelled or stuck over with laurel wreaths," engines that had "become very miserable failures in the attempt at something elegant."[27] Ever the classical purist, Clegg the aesthetician assumed that true elevance and grandeur lay in a strict attention to aesthetic principles derived from classical antiquity, based on correct proportion and simplicity of design.

Yet in his dual role as engineer, Clegg also gave a strong technical argument for simplicity in machine design, insisting that a well-proportioned engine had an inherent elegance that obviated the need for superfluous ornamentation. He argued that engine manufacturers were mistaken who thought of "*hiding* essential parts of a machine by adding useless ornament." Elegance was achieved through "stability without unnecessary weight, simplicity of form without meanness." A well-designed engine did not need any ornamental trappings to heighten its stature: it created "an elegant figure, having beauty of form without any attempt at grandeur."[28]

(Clegg's emphasis on design simplicity and his abhorrence of false grandeur again suggested the underlying concern with social class and pretentiousness seen in nineteenth-century British design criticism. As he chastised manufacturers, it would "not cost one farthing more" to create an elegant form "than it does to create a deformity, or to make a machine in such a manner as to raise disgust by its pretensions at display." While referring to machine aesthetics, his language also spoke of social pretensions: "Deformity decked out in finery is hideous," he wrote, adding, "'Pretend to nothing but what you are' will apply to man's handy work as well as to man himself, and to attempt to give beauty by polish, when the form is unsightly, must always be pretence.")[29]

Clegg's notion of simplicity did not imply that he advocated totally bare machine forms, nor was it an early formulation of an aesthetic of unadorned functionalism. Accepting his contemporaries' practice of architectural engine framing, he endorsed well-proportioned engine columns based on Roman Doric models and appropriately proportioned classical cornices, pedestals, and entablatures that retained the intent and character of antiquarian models. But as an engineer, he recognized that aesthetic choices were often necessarily subordinate to engineering requirements, and he attempted the difficult task of reconciling engineering needs with aesthetic propriety. As he argued, "the laws of statics will be found to combine with those of taste in the production of good design to so great an extent that they will appear dependent on each other."[30]

Clegg did recognize, however, that engineering requirements and aesthetic issues were often at odds, and as an engineer and classical purist he was intent on attempting to reconcile the two, insisting that if functional requirements necessitated making a gross distortion in classical proportions, the engineer must "eschew the fashion of architecturalizing machinery." As an example, he examined the appropriateness of Tuscan, Grecian Doric, and Roman Doric orders for engine design, noting that although the stark simplicity of short Tuscan columns would be best for machinists, the requirements of iron casting would result in aesthetic distortions, causing the Tuscan character to be lost. The best stylistic compromise, he concluded, would be Roman Doric, though he lamented that even "its dimensions must also frequently be made at variance with architectural rules."[31]

The engines being manufactured in the 1840s and earlier did indeed use Roman Doric and Grecian-order columns and were roofed with entablatures based on Roman Doric models, including triglyphs in the frieze. These engines appeared to have been modeled after the Roman Doric architecture seen in eighteenth-century pattern books such as that by Isaac Ware, which included examples of domestic structures influenced by Palladian designs.

To clarify the contrast between classical architecture and contemporary steam engines, Clegg included in his text a diagram of a column from the famed sixth-century Greek Doric temple at Paestum juxtaposed next to a column from a nineteenth-century house engine, with the engine's incorrect proportions carefully noted. Clegg was particularly critical of engines that were guilty of "distorted imitations"—Doric detailing marred by disproportionate columns and "mouldings disfigured under the hand of the draughtsman or pattern-maker"—and he illustrated his complaint by contrasting a column from the Parthenon in Athens with that of a cast-iron nineteenth-century marine steam engine that had a more slender column, narrower capital, and truncated entablature (fig. 60). (Eighty

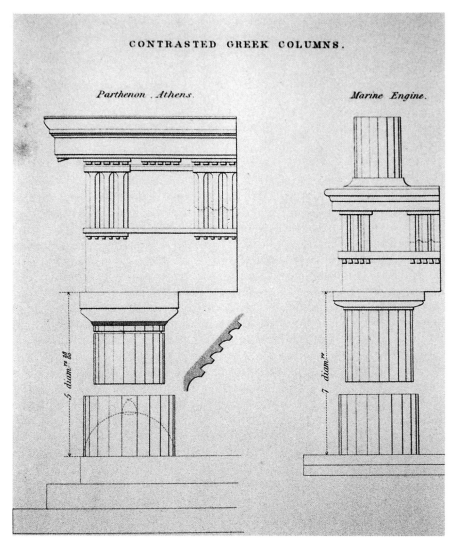

60 Contrasting Greek columns, Parthenon and marine engine. Samuel Clegg, *The Architecture of Machinery* (London, 1842).

years later, Le Corbusier, writing in *Towards a New Architecture* [1923], would also cite the Parthenon as a model for modern architecture and technological design, lauding its perfection and precision of parts as a paradigm for the machine age.)[32]

Intent on reforming contemporary practice in engine design, Clegg formulated a list of rules, which included stipulations about correct proportion. Here again he attempted to reconcile engineering needs with aesthetic purity and showed a willingness to sacrifice classical ornament if necessary: "A column should never have a less diameter than $\frac{1}{10}$th the height of its shaft. If essential that the support be of less strength than this rule will give, depart from the regular orders entirely, and employ some pleasing curve in the upper member."

Directly referring to contemporary uses of the Roman Doric order, he advised: "Triglyphs should be omitted generally, or except in instances where the columns are the correct distance apart, which seldom happens in machinery, because they must be placed not with reference to themselves, but to the parts of the machine they carry." Clegg concluded that any other architectural borrowings might "lead us into error and absurdity" and advised that "all we can do with 'framings,' beams, and cast iron work, which partake of no uniform figure, is to design them with judgment, good taste, and proper regard to those laws which regulate the actions of forces."[33]

Throughout *The Architecture of Machinery*, Clegg's concern was in correcting structural and aesthetic absurdities that he saw in contemporary engine design. One of his central targets was a version of engine design metonymy in which a single architectural part, frequently a classical column, was used to suggest the whole. By 1840, manufacturers were designing steam engines that made use of a single fluted classical column to support the frame, a practice that Clegg abhorred: "Although the proportions of this example are in accordance with those given by Athenians, it looks strange and out of place: there is no authority for a single column of this order having been found, the beauty of which depends on the general assemblage of many features rather than the contour of detached proportions."[34]

A second area of complaint in Clegg's work focused on engine designs that reflected the engineer's concern with structural stress and strength of materials. Although earlier he had argued that correct aesthetic taste was informed largely by imitating classical proportions, he also argued for a more specifically engineering aesthetic, one that defined elegance in terms of using a minimal amount of material to achieve the desired degree of strength: "beauty of proportion in machinery is sequent upon the use of those figures which possess the greatest strength with the least possible matter." Absurd results occurred when designers, in their quest to use minimal material, violated the need for structural stability: "It is painful to contemplate a huge mass of matter sustained upon such slender supports that they bend and totter under it." Equally ridiculous, Clegg contended, were engines that failed to fulfill the psychological necessity of creating the appearance of stability. He described one engine pedestal as being "meager, and totally devoid of that appearance of strength which a foundation ought to exhibit." Other pedestals were designed so poorly that they gave the impression of "figures on stilts."[35]

EVOLVING A DESIGN AESTHETIC

The aesthetic issues defined by Clegg in 1842 were early formulations of the broader concerns that would occupy the thoughts of design critics and engineers throughout the century as they continued to seek an appropriate style for emerging new machines. Clegg's arguments for a balancing of aesthetic issues and functional demands were soon taken up by British critics including M. Digby Wyatt in his *Journal of Design and Manufactures* article (1850) mocking and mapping the conflict between the ornament-happy "idealists" and the function-minded "utilitarians." Just as Samuel Clegg had complained about design metonymy and the use of a single cast-iron Doric column in steam engine design, the *Journal* derided the "use of a shaft for a lamp-post, with as few symptoms of an entablature." But while Clegg's treatise condoned the use of classical design elements as a means to enhance the elegance and stature of the steam engine, Wyatt, with a note of status consciousness, complained about the debasement produced when classicism was applied to quotidian uses, seen in "the bringing down of the dignified Greek Doric column to the situation of cast-iron balluster to a staircase or balcony."[36]

Undeterred by critics' debates about the proper uses of ornament, British and American manufacturers of industrial and domestic machines continued their practice of decorating cast-iron engine casings with historical design imagery and painting machines with ornamental patterns, a practice particularly popular in America after 1850.

Classicizing the Machine   199

61 Atwater sewing machine (1860). National Museum of American History, Smithsonian Institution, Washington, D.C.

The American-made Atwater sewing machine of 1860 was topped with a tiny silver neoclassical urn (fig. 61), and a West and Wilson machine (1859) was framed with a stocky cast-iron fluted column.

Americans from the 1850s through the 1880s were noted, if not notorious, for ornamenting their machines not only with classical and Gothic-style cast-iron frames but also with fanciful floral images, scrolls, and arabesques painted on machine casings. After initially introducing an unadorned sewing machine in 1851, American manufacturer Isaac Singer soon covered the cast-iron machine casings with a black japanned finish ornamented with stenciled or decaled flowers—a practice mimicked by Remington and other manufacturers and continuing into the next century.[37]

American engine and machine-tool manufacturers often painted their engines deep red or hunter green with stylized floral patterns in gold paint. The engines and machine tools manufactured by Putnam of Fitchburg, Massachusetts, in the 1860s were characteristically painted deep olive green striped with dark red and purple, a practice often derided by British observers but continued by Americans through the 1880s. Putnam's cast-iron planing machine of 1865 was ornamented with scrolls and leaf imagery and rested on four squat "turtle" feet.

The nineteenth-century practice of ornamenting engines, machine tools, and domestic machinery has been attributed to a variety of factors, including the century's love of ornament itself, the inheritance of an eighteenth-century tradition of ornamenting machine tools and scientific instruments designed for aristocracy, and, as John Kouwenhoven argued, the training in ornamental detailing received by early American carpenters and cabinetmakers who transferred their skills when producing woodworking machines.[38] Americans, it has also been argued, ornamented machines to assimilate them, to make them look artistic, and to reflect their honored status in national life.[39]

But there were also lingering suggestions that ornamenting American engines manufactured later in the century helped soften their association with danger. A small Tyson engine manufactured in Philadelphia for domestic use in 1881 was set in an ornamental vase adorned with the figure of a woman, and it rested on three legs or was hung on a wall bracket. The *Manufacturer and Builder* wrote that previous efforts at designing small engines for domestic use had produced machines that "were heavy, bulky, and expensive; or were unsafe on account of a liability to explode," but the Tyson engine was said to have solved these difficulties.[40] Cast as a decorative object for the home, the Tyson "vase" engine helped domesticate the machine's explosive potential.

There were continuing evidence, too, that ornamented machines reflected not only an important marketing strategy but also middle-class longings for respectability and status. The new machines were not just technological but also social artifacts, objects that some might consider vulgar interlopers, that had rather suddenly appeared on the social scene and, as though uncertain of their own status, had dressed up, as it were, for the occasion.

Such a theme of social aspiration pervaded the description given by John Leander Bishop in his *History of American Manufactures* (1868) of his visit to the Wheeler and Wilson Sewing Machine Company, where he viewed the manufacture of wooden sewing machine cases. Bishop's description approvingly detailed the upwardly mobile aspirations of both consumers and manufacturers, who had created cases "made of polished black walnut, mahogany, and sometimes of rosewood and other costly materials," cases that, when "made in the same way, and when elaborately ornamented, to correspond with the costly finish of the higher priced machines, form elegant articles for the boudoir." Bishop with apparent pride added that the Wheeler and Wilson salesroom on Broadway was "fitted up in a style of regal magnificence."[41]

While designers were ornamenting new machines in part to enhance their status, British commentators on occasion put forth the relatively radical proposal that plain designs were a surer sign of dignity and secure status. At midcentury, the debate about the status of new technologies was crystallized in reactions to the machinery displayed at London's Great Exhibition of the Industry of All Nations in 1851. Manufacturers drew heavy fire from British design critics including Henry Cole, who saw in many heavily ornamented industrial products a violation of the preferred design aesthetic balancing utility and ornament.

To bolster its position, Cole's *Journal* in 1851 reprinted a revealing exhibition review published in the London *Times* which found the misuse of ornament offensive: "The most serious violation of principle common to all nations is the negation of utility as paramount to ornament." In its analysis, the *Times* revealed the preoccupation with social pretentiousness that often shaped British discourse on industrial design. After denouncing the large number of lamps, candelabra, and chandeliers reflecting "atrocities of taste," the article praised the design of machinery:

> Some sections [in the exhibition], and especially that of machinery, feeling their pre-eminence secure and undoubted, have been content to be plain and unpretending in consequence of which they develop a high degree of artistic excellence. The most refined taste will gather pleasure and satisfaction from a survey of our machinery department.[42]

Unornamented machines were praised for their "truthfulness, perseverance, and severity," qualities helped create "a style of art at once natural and grand." The *Times* cited as a particular example machine tools made by British manufacturer Joseph Whitworth, who was to become known for his functional designs.

Machinery with a utilitarian design thus became, in the *Times*'s view, an exemplar of truly aristocratic sensibility: having a secure sense of status and legitimacy, such machinery could afford the dignity of being "plain and unpretending," evoking grandeur without the pretentious social trappings associated with tasteless ornament. But while finding plainness the appropriate aesthetic for machinery, the *Times* did not go so far as to recommend functional plainness for domestic products, and it included in its review an important disclaimer: "We do not for a moment contend that the unbending precision which produces such great results . . . would be equally applicable to the manufactured products" for "our everyday and domestic wants and comforts," but "we cannot with impunity attempt to recall defunct or foreign styles of ornament."[43]

While design critics pondered the proper role of ornament applied to machinery, beam engine designers after midcentury were still reluctant to address the issue of aesthetics. Oliver Byrne's treatise described the illustrated single-column Corinthian-design steam engine in vague terms of engineering minimalism, ascribing its popularity not only to its operational steadiness but also to the "simplicity and elegance of its general arrangement."[44] *Engineering*, a journal first published in Britain in 1866 and one of the first journals devoted to the emergent profession of engineering, included in its initial volume an article "Form in Design" which complained, "Whatever engineers may believe respecting taste in design, it is very seldom that any reference is made to it in their reports." It added, "Many engineers, accustomed to think in steam or iron have brought themselves to despise all professors of taste," and now show "less regard for taste than formerly." Ending on a note of despair, the article concluded that too many engineering works were "hopelessly ugly."[45]

Though *Engineering* complained about engineers' lack of interest in evolving an engineering aesthetic, its own reports generally continued to ignore the aesthetic appearance of ornamented engines. In 1867 the journal described an engine for rolling mills designed by Hargreaves and Company with Doric columns and entablature but never mentioned the engine's architectural frame. There was, though, occasional praise for neoclassical frames in engine design. Describing a gasholder at the Fulham Gasworks in England, *Engineering* in 1868 cited the elaborate composite Corinthian columns with ornamental entablature, calling the gasholder probably "the handsomest structure of its kind which has as yet been erected." In its design, the engineer Thomas N. Kirkham had paid "no less attention to the artistic than to the mechanical details," resulting in "a work of not only engineering interest, but also very considerable architectural merit."[46]

If British and American engineers, in the process of professionalizing, were relatively silent about engineering aesthetics, there were some occasional early efforts at formulating an aesthetic for machine design. In its first volume, America's *Appleton's Mechanics' Magazine and Engineers' Journal* (1851) included the article "Men and Things Mechanical"—a remarkably bold early proclamation of functional machine beauty. But while championing the machine, *Appleton's* reflected the century's ambivalence by assuming that it was the inner elegance, not the outer appearance, of machinery that was beautiful.

With the inflated rhetoric of a manifesto, the magazine began by proclaiming that machinery has an inherent artistic quality too long ignored: "We do condemn and denounce, and earnestly deprecate the shamefully prevalent neglect of things *mechanical* by those who pay a willing homage to things picturesque, poetical, or philosophical."[47] Using a ship as an analogue for machinery, the article assumed an inherent dichotomy between the ungainly exterior of the machine and its inner, conceptual beauty:

> There is in this world . . . a beauty for the mind as well as a beauty for the eye. The Creator does not always give these good gifts in conjunction. There is one glory of the star, another glory of the flower, . . . and another glory of the *machine*; and because the star, and the flower, and the diamond, are beautiful to the eye, we do not deny to them the inner and spiritual beauty which they offer to the mind. Why, then, should we deny the inner and spiritual beauty of an artificial machine, because the organic body wherewith it is clothed is rough and unsightly, because it emits no tender perfume, and reflects no brilliant ray?[48]

But while the journal praised the machine's inner beauty, it left no doubt that machine exteriors with their "rough" and "unsightly" bodies were decidedly inelegant. Making an analogy to paintings in a gallery, the article noted that with the mere addition of an appropriate ornamental frame, the humble machine would fully reveal its inherent beauty: "Here is a picture on coarse canvas with a common frame, discolored with time and dust; let us wash away the stains, and put a gilt frame upon it, and it will stand the loveliest picture in the gallery." But with ambivalence, *Appleton's* concluded by making another apparent tribute to the inner beauty of well-designed unadorned machines, a beauty without the need for ornament: "possessed of rich, intrinsic beauty, emanating from the hand of *a master*," the machine "disdains the adulteries of outward show."[49]

With piety and technological chauvinism, *Appleton's* editors thus sang their praises of the machine with almost religious fervor, yet ultimately they could smell no "tender perfumes," and stopped short of seeing any aesthetic beauty in functional design. But at midcentury, it was the American sculptor Horatio Greenough who, however unorthodox his position, was to argue for the inherent beauty and elegance of functional technological forms. In an age permeated with stylistic quotations from classical Greek design, Greenough turned to the aesthetic values of classical Greece to bolster his belief in functional design. Greenough, who graduated from Harvard, had studied art in Rome and was best known for his neoclassical sculptures, particularly a controversial marble version of George Washington as Zeus scantily clothed in a classical drapery. In an era devoted to ornamentation, Greenough argued that the unornamented machine, when designed according to Greek principles and functional criteria, had its own inherent aesthetic elegance.

Greenough's protofunctionalist aesthetic was voiced in his essays and newspaper articles, which were often published using the pseudonym Horace Bender and were later collected in two separate volumes: *The Travels, Observations, and Experiences of a Yankee Stonecutter* (New York, 1852), published only four months before his death at age forty-seven, and *A Memorial of Horatio Greenough*, published posthumously by Putnam in 1853. It was here that the sculptor associated the unadorned, functional design of machinery with Greek culture and classical principles, a rhetorical coupling that would become a central tenet in the modernist design theory of Le Corbusier and the machine art aesthetic that was to permeate America's design consciousness in the 1920s and 1930s.

In the midst of the American neoclassical architectural revival of the 1840s and 1850s, Greenough drew on references to ancient Greece to urge architectural reform as he championed classical proportion, functionalism, and purity of form while also praising the functional simplicity of the machine as a paradigm of elegant design. Greenough's writings castigated American architects for their unthinking imitation of classical architectural models and their use of inessential, pretentious Greek ornament, citing the sailing ship and the wagon as models of functional design:

> The men who have reduced locomotion to its simplest elements, in the trotting wagon and the yacht *America*, are nearer to Athens at this moment than they who would bend the Greek temple to every use. I contend for Greek principles, not Greek things. If a flat sail goes nearest wind, a bellying sail, though picturesque, must be given up.[50]

Greenough's association of classical Greek principles with the functional simplicity of the machine was again suggested in his essay "American Architecture," where he derided Greek architectural facades and urged, "Let us learn principles, not copy shapes."[51] Criticizing American architectural imitation of Greek temples, he urged American architects to strive for simplicity and subordinate parts to the whole, and cited the "newly invented machine" as a model of adapting form to function, noting that as superfluous weight is shaken off, "the straggling and cumbersome machine becomes the compact, effective and beautiful engine."[52]

Amidst Greenough's despair over omnipresent classical Greek ornamentation, and *Appleton's* foray into the idea of unadulterated machine beauty, manufacturers continued to produce ornamented engines, with the production of British and American classicized steam engines and small domestic machines reaching its peak in the 1860s. British six-column beam engines with fluted columns were built through the 1870s and beyond, and Egyptian and Gothic-style engines also continued to be built after midcentury.

One of the reasons for the persistence of architecturalized engine frames was that the practice helped transform steam engines into imposing symbols of national pride, and helped heighten the prestige of steam technology itself. The gilded and canopied steam engine *Southern Belle*, used to power the machinery exhibits at the New York Crystal Palace in 1853, became a fitting emblem of the country's pride in its technological achievements, and the towering Gothic arches framing the pumping engine designed by William McAlpine in 1857 lent status and stature to the engine room of the U.S. Navy Yard Dry Dock in Brooklyn.

In a century preoccupied with social progress, ornamented steam engines also came to signify achievements in public sanitation: sewage works and waterworks were powered by "temples of steam." At the British Whitacre Pumping Station in Staffordshire, the inverted compound beam engines built in 1885 by James Watt and Company were given grandeur with imperial Napoleonic columns and corbels designed as gilded eagles. Americans also used ornament to transform public utilities into monuments of civic pride. With its echoes of classical Greece, the Philadelphia Gas Works, approved for construction in 1835, was a gargantuan, eight-acre testament of faith to a new technology. Among the eleven gasholders, one of the largest was 140 feet in diameter, consisting of 144 columns all standing four tiers high and designed to represent different classical orders—Tuscan, Doric, Ionic, and Corinthian—with the entire superstructure rising 74 feet high.

Receiving praise in *Gleason's* magazine in 1853, the gasworks was described as "one of the chief attractions of the city," though the magazine punningly noted that the "unen*lightened* corporation of Philadelphia" had initially "shrunk from innovation" and had been cautious about spending money on a technology it "regarded as an unsafe means of yielding artificial light." The structure's classical imagery, with its associations of tradition and security, may well have been a particularly wise design choice for this potentially explosive technology.[53]

But monumentality and grandeur could also be achieved without ornate historical trappings: the Corliss engine that powered the exhibits in Machinery Hall at the Philadelphia Centennial Exhibition of 1876 rose thirty-nine feet with massive splendor and devoid of any decorative detailing. But while the Corliss engine won praises, America's decoratively painted machines were at times ridiculed by Britain's observers, including the journal *Engineering*, which complained about "grotesque ornaments and gaudy colours."[54] Though a cast-iron portable steam engine such as "The Iron Slave" manufactured by L. Sweet in New York in 1870 was ornamented with scrolls, leafage, and fruit and though Westinghouse engines of 1882 still had ornamental fluting, American manufacturers by the 1880s showed signs of favoring design simplicity by turning away from blue or green engines with maroon stripes to lead-color engines with unornamented, functional designs.

There were indications, too, that unornamented machine designs were being viewed by engineers as a welcome aesthetic. In his autobiography, Anglo-American Hiram Maxim, inventor of steam engines and best known for the Maxim machine gun, described his visit to his uncle Levi Stevens's Putnam machine works in Fitchburg, Massachusetts, where engine stripes and ornamental painting were giving way to gray paint. Viewing a Putnam steam engine on display at the American Institute Fair in Boston, Maxim praised the gray engine as "much better than the usual gaudy greens, purples and reds." "It had," he added, "a certain chaste and elegant appearance."[55]

Observing the country's changing design practices, American engineers and machinists, in their infrequent evaluations of aesthetic practices, showed signs of increasing self-criticism and self-consciousness. *The American Machinist* in its article "Machine Aesthetics" (1882) complained that designs of twenty-five years past had been more elegant. In the case of the engine lathe, "more effort was made in the direction of structural beauty, simply for the purpose of presenting pleasing forms to the eye, by the leading manufacturers then, than is made by the same class today." Contemporary engine lathes "have gradually taken on a utilitarian form, to the elimination of anything that could be construed as a distinct attempt at structural beauty."[56]

But while lamenting that current machines lacked the fine finish of former manufacturers, the *American Machinist* presented a vision of the inherent beauty of functional forms. It would be a mistake, the journal averred, to assume that "there was nothing of the aesthetic" about current machine design: "It has simply changed conditions, so that its appreciation appeals to other senses as well as the sense of seeing. Forms are no longer considered beautiful because they merely please the eye, but they please the eye because they appeal to a more substantial consideration of their purposes." The magazine's aesthetic emerged as one in which the geometries of functionalism became central: "the making of cylindrical parts *cylindrical*, and flat parts *flat*" was "considered a part of the beautiful in machinery."[57]

A revealing end-of-century effort to articulate an engineering aesthetic was presented in an article "Aesthetics in Machine Design" by the mechanical engineer John H. Barr and published in the American engineering journal *Cassier's Magazine* in 1892. After acknowledging that "strictly utilitarian elements"—the costs of construction and maintenance, durability, convenience in operation, the availability and appropriateness of materials—were "of the first importance" in industrial design, Barr asserted that aesthetic issues, although secondary, should probably be included among the prime requisites. Calling the neglect of appearance "inexcusable," the author, however, made clear his aversion to the contemporary practice of machine ornamentation:

> This is not to be considered as a plea for floral decorations, for Corinthian columns, for moldings and cornices, or for red paint and yellow stripes, not even for an excess of less objectionable ornamentation, such as polished brass and drawfiled surfaces. These latter may be very effective in the proper place, but no array of them will compensate for the absence of easy, natural lines and harmonious proportions.[58]

Barr's machine aesthetic, like Samuel Clegg's fifty years earlier, placed an emphasis on "simplicity and harmony." He praised recent attempts to minimize architectural detailing in machine design. Citing the "wonderful progress" reflected in contemporary machine trade catalogs, he noted, "The designs of the more progressive builders are all characterized by outlines of extreme simplicity, with no unnecessary moldings, ledges, cornices, etc." While *Appleton's Mechanics' Magazine* of 1851 saw the need for "gilded frames" on bare machines, Barr's engineering aesthetic of 1892 criticized ornament and presented a more confident affirmation of the legitimacy of unadorned machine forms:

> As utility is *the* consideration in constructing a machine, simplicity and harmony should be the aims of the designer in striving for the best and most appropriate appearance. Ornament for the sake of ornament is to be rigidly avoided. It is very bad design, like disagreeable medicine, that needs a sugar coating.[59]

Rejecting the century's practice of joining classical ornamentation and the machine, Barr's article concluded with a dismissal of classicized engine rooms and added further praise for functional simplicity. Contrasting two Chicago waterworks stations, he noted "the interior of the North-Side Station is filled with specimens of Greek architecture, reminding one more of a classical museum than of an engine room: yet these same engines were looked upon, not many years since, as engineering achievements of a very considerable note." Far more satisfactory, in Barr's view, was Chicago's Harrison Street Waterworks Station on the city's West Side, where the triple expansion engines designed "by that king of designers," Edwin Reynolds, were the essence of utilitarian elegance: "Every line of them is pleasing; every feature is an essential part; the whole is a magnificent exemplification of the majesty of machinery."[60]

For all of Barr's disdain of Greek architectural ornament, classical ornament and technology continued to be linked through the end of the century. The buildings for Chicago's Columbian Exposition (1893), with its celebration of new technologies, were designed using a classical idiom which countered the late Victorian infatuation with neo-Gothic design and helped launch another neoclassical architectural revival in the United States. But increasingly, architects and design critics seemed to be echoing Greenough's call for "Greek principles, not Greek things."

The commingling of two nineteenth-century preoccupations—neoclassicism and the search for an aesthetic style befitting emergent industrial societies—became increasingly apparent at the end of the century among designers and design theorists reacting against the excesses of art nouveau as well as the stylistic array of the classical, medieval, Renaissance, and rococo imagery that had ornamented nineteenth-century industrial designs. The search for design coherence and the continuing allure of classical ideals were shaped by a complex array of nineteenth-century design currents, including Winckelmann's writing on neoclassicism widely read by design theorists, the impact of the stripped classical designs of Prussian architect Karl Friedrich Schinkel, the quest for design reform by William Morris and the British Arts and Crafts movement, and the writings of French architect Viollet-le-Duc in the 1860s and 1870s urging engineering rationalism and a new style for an industrial era.

During this period, when designers were ceasing to architecturalize steam engines with Doric columns and entablatures, design theorists sought to transform industrial design through an aesthetic that once again turned to ancient Greece, at first as a model for functional design simplicity and later as an inspiration for the abstracted, unadorned geometries of modernist design. Rhetorical allusions to ancient Greece were particularly evident in the essays of Austrian architect Adolf Loos (1870–1933), who saw in the great achievements of ancient culture an apt analogy for the brilliant functional works of contemporary engineers. Through his inflated, enthused rhetoric, Loos presented yet another instance of alluding to classicism to lend status and stature to modern technology.

Loos, in his essays published from 1897 to 1900 and collected in the volume *Ins Leere gesprochen* (Spoken into the Void, 1921), was an avid admirer of modern machinery and things Greek. In his essay "Glass and Clay," he lauded Greek vases as objects "made so practical that they could not be made any more practical, then they called it beautiful." Asking, "Are there still people today who work as the Greeks worked?" he answered emphatically: "Yes indeed! The English do so as a nation, the engineers as a profession. The English and the engineers are our Greeks. It is from them that we acquire our culture.... They are the consummate men of the nineteenth-century."[61]

Equating Greek functional beauty and modern machine design, Loos wrote in "Glass and Clay": "The Greek vases are beautiful, as beautiful as a machine, as beautiful as a bicycle." And in his "Review of the Arts and Crafts" (1898), he praised the elegance of the enormously popular, industrially produced bentwood Thonet chairs again in terms of fifth-century Greece: "Look at the Thonet chair! Without decoration ... is it not born out of the same spirit as the

Greek chair with its curved feet and its backrest? Look at the bicycle! Does the spirit of Pericles' Athens not waft through its forms?"[62] Loos's own version of classical design principles was reflected in the bare abstracted geometric volumes of his concrete houses at the beginning of the next century, particularly in the clarity and stark rectilinear forms of his Steiner House (Vienna, 1910).

Through their use of classical ornamentation, nineteenth-century steam engine designers had discovered a means to associate the new technology with a historic model of dignity and stability: the correct Doric proportions prescribed by theorists such as Samuel Clegg, if imitated with care and sensitivity, balancing a concern for classical stylistic conventions with the needs of engineering, promised to produce machines endowed with the status, grandeur, and elegance of the ancient classical models—machines, like the ancient classical temples, endowed also with an aura of timeless beauty, permanence, and stability.

By the end of the century, however, the technological aesthetic that produced the steam engine as classical temple was being replaced by new representations of the machine age: Loos's bare, concrete, geometric architectural volumes became harbingers of the twentieth century's developing modernist aesthetic, which saw functional (in theory if not in practice) and unornamented forms as the most appropriate style for industrial societies—a style which, in embracing the purity and timelessness of classicism, provided a way to transcend the stylistic changes and confusions of the previous century.

This newest version of classicism with its Apollonian promise of a rational order and restraint would again provide a needed semblance of timelessness and stability amidst the twentieth century's own confrontations with explosive technological and social change. After a century of efforts by manufacturers, designers, and engineers to legitimize new technologies, to reconcile the ruptures and ease the impact of broken frames, the plain and proud, upstart machine—now elevated to the status of an icon—had at last arrived.

Afterword

Into the Twentieth Century

The technological landscape has changed greatly since Philip de Loutherbourg painted *Coalbrookdale by Night* at the beginning of the nineteenth century and John Ruskin, a half-century later, elegiacally warned of "the darkness that broods over the blast of the Furnace and the rolling of the Wheel." With its fiery image of iron foundry flames lighting the sky at night, de Loutherbourg's painting summoned up the early nineteenth-century view of technology as towering and potentially threatening, both terrible and sublime. The smoke pouring from Coalbrookdale's furnaces signified industrial progress, but it also threatened to obscure the moon—to challenge, if not overshadow, nature itself.

But in the early decades of the twentieth century, technology seemed less threatening to artists who captured the electrical excitement of a new technological era. Electric lights were brightening American and European cities, and in *Street Light* (1909), painted by Italian Futurist Giacomo Balla, the fracturing that haunted nineteenth-century images of technology and the radiating lines that once suggested lethal explosions were now transformed into small chevrons of red and yellow light radiating outward, suggesting an age enthralled with new technologies. The rays of light cutting through the darkness in Balla's painting coexist with the crescent moon: technology vies with nature rather than threatening it, offering a powerful new means of illuminating the night.

With their images of jarring explosions, nineteenth-century artists shattered the myth of technology's easeful entry into the social terrain. But Balla's electric light in early twentieth-century art suggested a world grown accustomed to the presence of new technologies. Technology, no longer jostling for a comfortable place in the landscape, now radiates outward, pushing the boundaries of the pictorial frame, extending notions of the proper subject of art itself.

During the 1870s, British and American engineers had begun to engage in self-evaluation, questioning the need for ornament and tentatively arguing for the inherent beauty of machines. In 1876, the year of America's centennial celebration, Chicago's *Engineering News* looked forward to the public's advancing awareness of "graceful form and design."[1] This increasingly self-conscious concern with improved design and the dignity inherent in machines paralleled the arrival of machinery as a new, privileged subject in early twentieth-century art.

Philadelphia-born artist Morton Livingston Schamberg was among the first to focus on the clean lines and classic simplicity of machines. *Mechanical Abstraction* (1916) gave the geometries of machine parts an air of quiet dignity and fluid grace, and the artist's painting *Telephone* (1916) became an elegantly choreographed dance of cool blue cylinders and cones (fig. 62). (Schamberg's paintings of the telephone and dynamos were bold and striking images without symbolic or satiric overtones, yet he was not always so straight-faced in his view of technology. Mocking widespread admiration for America's modern plumbing, the artist mounted a plumbing trap on a miter box and entitled his sculpture *God*.)

Charles Sheeler's Precisionist paintings presented some of America's most elegantly distilled and evocative images of machines in an industrial world. In *Classic Landscape* industrial silos and smokestacks emerge as strong, stark structures cast in a clear, sharp light. Alfred Barr, Jr., of New York's Museum of Modern Art argued in 1934 that machines are more "beautiful as objects when they are still," and in Sheeler's painting, industry is transformed into a vision of enduring, classical calm.[2] While nineteenth-century designers had dignified new steam engines with classical ornament, Sheeler's industrial and machine images have a classical simplicity, their pristine forms needing no classical ornament to assure their prestige.

Classic Landscape placed machine-age technology in a timeless world, but this was a world not about to stand still: aided by new technologies, artists continued to interpret the landscape anew. Almost fifty years later, computer-generated images were redefining

62 Morton Livingston Schamberg, *Telephone* (1916). Oil on canvas. Columbus Museum of Art, Ohio, Gift of Ferdinand Howald.

the landscape, creating sleek, simulated city scenes and craggy fractal mountains set against violet skies.³ Nineteenth-century critics had been wary of the imitative arts, seeing them as a challenge to hand-created, original works of art. But in the world of computer graphics and computer animation, synthetic landscapes have in a sense become even more subversive, confounding and supplanting the real with its electronic facsimile.

Twentieth-century artists continued to reinterpret technologies in other new ways as well, exploring the issues of fracturing and integration which had troubled and tantalized artists during the century before. Artists' subjects in the age of steam—mechanical speed, fractured machines, explosions, the increasingly depersonalized human psyche—appeared in twentieth-century art tracing the impact of technology in the modern, and postmodern, age.

In their satires of technology, nineteenth-century British artists saw the humor and danger of people becoming machinelike in an increasingly mechanized world. Robert Seymour's fantasies of people walking with steam-powered legs and artists' images of figures surrounded by encircling wheels highlighted the problematic nature of human identities becoming ever more symbiotically connected to machines. Twentieth-century artists continued to be both fearful and fascinated by the prospect of mechanized human figures, seeing in them the very embodiment of the modern era: wrote German Bauhaus artist Oskar Schlemmer in 1922, "Life has become so mechanized, thanks to machines and technology . . . that we are intensely aware of man as a machine and the body as mechanism."[4]

During the early decades of the century, the European avant-garde presented abstracted, admiring views of mechanistic human figures, seen in Schlemmer's own Bauhaus theater costumes and smoothly polished metal figures with featureless faces. French artist Fernand Léger was famous for his paintings of tubular human figures whose bodies had the polished sheen of machines. And in an early use of celluloid, Russian-born artist Antoine Pevsner fabricated his figural *Torso* (1924–1926) as interpenetrating plastic planes engineered with machine-part precision.

But artists of this century have also considered the prospect of human automatons a chilling, nightmarish possibility. Following the horrors of World War I, a highly industrialized war, German artist George Grosz created caustic views of citizens who had become mindless cogs in an impersonal and impoverished machine culture. In an untitled painting of 1920, a solitary human figure stands in a factory zone, a person whose head has been reduced to a faceless sphere and whose cylindrical arms have no hands. In an age of advanced technology, Grosz suggests, individual faces have become irrelevant and hands obsolete.

Satirizing the absurdities of the war and widespread machine worship, European Dada artists spoofed the impact of machine mania on emotional and erotic lives. Spanish artist Francis Picabia's *Machine Tournez Vite* (Machine Turn Quickly, 1916) presented two interlocking gears, discreetly labeled male and female, as a copulating couple with all warmth and sensuality removed from the scene. The artificially created monster in Mary Shelley's *Frankenstein* was reenvisioned by twentieth-century Dada artists who depicted a world of artificial conception and impersonal mechanohumans—a world of Marcel Duchamp's bachelor machines.

During the 1960s and 1970s, American Pop Art produced its own witty indictments of an impersonal, highly mechanized society where human identity and technology had become intimately and even indelibly fused. James Rosenquist's billboard-like paintings of lipstick and machine cogs, echoing Dada art, satirically linked eros and technology, while Andy Warhol courted an impersonal art style and twitted American culture with his deadpan insistence, "I am a Machine."

In the 1980s, New York artist Nancy Burson turned to contemporary computer technologies to challenge definitions of human identity itself, producing simulated images of the human face. Burson's composite photographs made with video camera and computer manipulations fused the faces of several people, creating hyperreal images of people who did not, and could not, exist. Her series of 1982 and 1984 merged the faces of American film stars, and her *Warhead* series (1982) combined the faces of political leaders in a nuclear age, producing a single hybrid face of a human war machine. Probing the ambiguities of female identity, in 1989 she fused the face of a woman and a mannequin, creating a startling android-like face with piercing woman's eyes.[5]

In nineteenth-century caricatures, people whose bodies were overtly machinelike suggested the impact of a world dominated by new steam inventions. Burson's composite photographs eerily evoke the problematic nature of human identity in a world dominated by electronic media and electronic images, a world where uniqueness disappears and individuality is imperceptibly lost. Denying their computer-manipulated, technological nature, these synthetic faces often appear genuine, blurring the difference between the artificial and the real. Like Frankenstein's monster, they take on a nightmarish life of their own.[6]

It has been not only the danger of depersonalization that has troubled artists of the past two centuries but also the more deadly possibility of self-destruction brought about by explosive nuclear technologies. While nineteenth-century artists brought to the surface fundamental fears of fractured human psyches and broken mechanical frames, twentieth-century artists were haunted by the devastating prospect of nuclear catastrophe—the most lethal version of the shattered earthly frame. For some artists, this threat was a virtual inevitability: as American artist Jenny Holzer, in collaboration with Lady Pink, tautly titled her painting of irradiated humans glowing red in a postnuclear world, *You Are Trapped on the Earth So You Will Explode* (1984).[7]

Taking a more sardonic view of technological destructiveness, French artist Jean Tinguely, best known for his "meta-mechanic" sculptures, created his *Homage to New York*—a conceptual sculpture designed to self-destruct in the courtyard of New York's Museum of Modern Art on March 17, 1960. Tinguely's sculpture was made of machine motors, bicycle wheels, an old piano, and a radio, and incorporated an addressograph and odd bits of old mechanical parts, including some found in the Newark city dump.

On March 17 at the assigned moment of destruction, fifteen motors sent soft-drink bottles crashing while a piano played, typewriters typed, and the addressograph worked as a percussion instrument. In a dramatic finale, the machine set itself on fire by pouring gasoline on the piano, but when the self-acting fire extinguisher failed to activate, the New York City fire department was called in, providing a final ironic commentary on the lethal quality, and sometime failures, of modern technology.[8]

Twentieth-century artists have also continued the nineteenth-century artists' fascination with mechanical motion and speed. While J. M. W. Turner's painting *Rain, Steam and Speed—The Great Western Railway* took an admiring view of England's Great Western Railway, other artists during the century feared that new technologies were apt to speed out of control. But in the early twentieth century, Italian and Russian Futurist artists relished the idea of speeding automobiles and trains, seeing in them the spirit of a radically new age. In their manifesto of 1910, the Italian Futurist painters demanded a massive reorientation of attitudes, a sweeping aside of the old subjects of art—landscapes, nudes, classical myths, historical events—in favor of images of new technologies: "all subjects previously used must be swept aside to express our whirling life of steel, of pride, of fever, of headlong speed."[9]

While the Italian Futurist artists, as seen in Balla's paintings, were intent on painting fractured images of speeding automobiles moving in time and space, avant-garde champions of the machine age were also turning to pictorial images of clocks with whirling hands to celebrate their sense that time itself was speeding up. In her painting *The Clock* (1910), Russian Futurist artist Natalia Goncharova superimposed a gold-colored spoked wheel over a large cobalt blue clock, with jagged lightning-like lines adding electrical crackle and sizzle to her image of rotary motion.[10]

A small number of American and British artists, including Joseph Stella and Britain's Christopher Nevinson, caught the Futurists' fervor, but some of the twentieth century's most striking emblems of speed appeared in American industrial design, where the adulation of airplane speed and styling became a central feature of the Streamlined Moderne.[11] As Sheldon Cheney wrote in *Art and the Machine* (1936), streamlining was an "aesthetic stylemark and a symbol of twentieth-century machine-age speed, precision, and efficiency"—a symbol "borrowed from airplanes and made to compel the eye anew."[12]

Inspired by the aerodynamic engineering of the Douglas DC-1, -2, and -3 aircraft of the 1930s with their contoured bodies and tapered wings, American designers including Raymond Loewy applied the stylistic features of a teardrop-shape, tapered body and chrome speed strips to railroad locomotives, automobiles, toasters, staplers, and even Loewy's 1934 design for a streamlined pencil sharpener. Streamlining became a major marketing device in a country infatuated with speed.[13]

But amid these celebrations of motion and speed, the old nineteenth-century dialectic was still there: the wish to stabilize a world of speeding machines in a framework of classical order and calm. Charles Sheeler's painting *Rolling Power* (1939, fig. 63), according to the editors of *Fortune* magazine, was inspired by a Hudson-type locomotive designed to pull New York Central trains at speeds of more than one hundred miles an hour—but in Sheeler's classicized view, the image is one of stasis, not speed.[14] Seen in bright light and free of any grease, grime, or rust, the locomotive's drive wheels and crankshafts become a marvel of technological power and elegance in a timeless world.

The classicized machine aesthetic in Sheeler's paintings was, in a sense, a continuation of the design dialogues begun in the nineteenth century. Samuel Clegg's treatise on machine design written in 1842 had explored suitable ways to join classical ornament and functional design, while America's Horatio Greenough at midcentury had ar-

63 Charles Sheeler, *Rolling Power* (1939). Oil on canvas. Smith College Museum of Art, Northampton, Mass.

gued in favor of "Greek principles, not Greek things." Reinterpreting the pairing of classicism and the machine, early twentieth-century modernist designers celebrated the abstract beauty of machine forms and saw in classical simplicity an aesthetic well suited to the modern age. The catalog for the Machine-Age Exposition held in New York in 1927 boldly proclaimed "the Machine and Mechanical elements as new symbols of aesthetic inspiration," and at New York's Machine Art show held at the Museum of Modern Art in 1934, stainless steel ball bearings, aluminum house pans, and ship propellers appeared as emblems of machine age functionalism and classical design (fig. 64).[15]

In his catalog essay for the 1934 Machine Art exhibit, Alfred Barr, Jr., with the stern voice of a machine age prophet, warned of the "treacherous wilderness of industrial and commercial civilization," where "on every hand machines literally multiply our difficulties and point to our own doom." The search for improved industrial design, Barr argued, could help heal this malaise: "we must assimilate the machine aesthetically as well as economically. Not only must we bind Frankenstein—but we must make him more beautiful."[16]

For early modernist designers, the route to assimilating and taming technology lay in unornamented, classicized design. Barr's catalog quoted approvingly from Plato's *Philebus*, which praised the timeless beauty of geometric forms: "By beauty of shapes . . . I mean straight lines and circles, and shapes, plane or solid, made from them

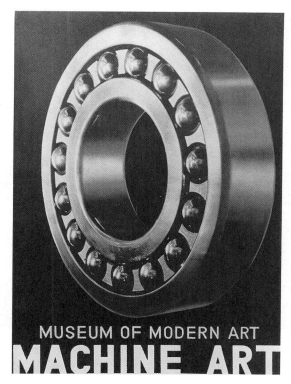

64 *Machine Art*, exhibition catalog cover, Museum of Modern Art, New York (1934). Collection of the author.

by lathe, ruler, and square. These are not, like other things, beautiful relatively, but always and absolutely."[17] The spare, classicized geometries and machine-part precision that had become the dominant aesthetic of modernist architecture were seen in Peter Behrens's AEG Turbine Factory (1908) in Berlin, a classical temple adapted for an industrial age, and the International Style architecture of Mies van der Rohe and Philip Johnson, whose New York Seagram Building (1957) became a monument of modernist, classically elegant glass-and-steel austerity.

Searching for an aesthetic appropriate for new technologies, nineteenth-century architects and designers of railroad stations and steam engines often turned to a form of classical metonymy: using a single element of neoclassical ornament to evoke a whole range of cultural and architectural associations. In the closing decades of

65 Charles Moore, Piazza d'Italia fountain, New Orleans (1978). Photograph: Norman McGrath, © 1978.

the twentieth century, postmodern designers and architects—now searching for an aesthetic appropriate for a highly technological, electronic age—revived the pairing of classical ornament and machine imagery, and again turned to design metonymy: classical stylistic quotations appearing as a thin veneer.

American architect Charles Moore's Piazza d'Italia (1978, fig. 65), dedicated to New Orleans's Italian community, was designed as a setting for festivals and celebrations. The piazza embodied what American architect Robert A. M. Stern has called "Ironic Classicism," fusing the materials of modern industry with sardonic references to the classical orders of antiquity. The piazza's planar classical temple fronts consisted of a series of semicircular colonnades that parodied the classical orders: Ionic columns with stainless steel volutes, Doric entablatures with cutout metopes that sprayed water (called "wetopes" by Moore), Corinthian and Ionic columns boldly illuminated by red and blue bands of neon light (what Moore puckishly called the "Delicatessen Order").[18]

The witty fusion of classical ornament and industrial materials in Moore's postmodern piazza extended the nineteenth century's efforts to achieve a rapprochement with contemporary technology while also presenting a critique of and tribute to the modernist aesthetic that dominated much of twentieth-century design. Through their use of classical ornament, postmodern architects of the 1970s and 1980s sought to bring a sense of warmth and historical continuity to the austere architectural icons of twentieth-century modernist and late modernist design.[19]

Moore's parodies of classicism also playfully extended the nineteenth century's often sober dialogue with the imitative arts. In *The Seven Lamps of Architecture* (1849), John Ruskin had denounced the deceitfulness of the imitative arts that covered base metals with a thin veneer. To Ruskin, these imitations were nothing but an "imposition, a vulgarity, an impertinence, and a sin." But the Gorham Company, which manufactured ornate electroplated wares, proudly proclaimed that their machine-made imitations were virtually indistinguishable from the real thing. Engaging in technological chauvinism, American cast-iron manufacturers even declared their imitations superior to handmade originals.

Reflecting a changed cultural climate, late twentieth-century architects and designers were no longer using ornament and imitation to dignify new technologies. Instead, Moore's neon-lit columns and Ionic capitals with stainless steel volutes sassily celebrated themselves as parodies, which became a source of their delight. By fusing classical ornament and twentieth-century technology, postmodernists were also, in a sense, continuing the quest begun in the nineteenth century: the search for order and tradition in a world still feeling the tremors of technological change.

Notes

INTRODUCTION

1. Much has been written about Sheeler and Precisionism. Martin Friedman's essay in the exhibition catalog *The Precisionist View in American Art* (Minneapolis: Walker Art Center, 1960) is a useful early commentary. A recent retrospective of Charles Sheeler's work was presented at Boston's Museum of Fine Arts. See the exhibit's two catalogs: Carol Troyen and Erica E. Hirschler, *Charles Sheeler: Paintings and Drawings* (Boston: Museum of Fine Arts, 1987), and Theodore E. Stebbins, Jr., and Norman Keyes, Jr., *Charles Sheeler: The Photographs* (Boston: Museum of Fine Arts, 1987).

2. Sheldon Cheney and Martha Candler Cheney, *Art and the Machine: An Account of Industrial Design in 20th Century America* (New York: McGraw Hill, 1936), 24.

3. Useful studies of technological innovation during the British industrial revolution include David S. Landes, *The Unbound Prometheus: Technological Change and Industrial Development in Western Europe from 1750 to the Present* (Cambridge: Cambridge Univ. Press, 1969), and Paul Mantoux, *The Industrial Revolution in the Eighteenth Century: An Outline of the Beginnings of the Modern Factory System in England*, rev. ed. (1961; reprint, Chicago: Univ. of Chicago Press, 1983).

4. Charles Francis Adams, *Notes on Railroad Accidents* (New York: Charles Putnam's Sons, 1879), 233.

5. Erving Goffman, *Frame Analysis: An Essay on the Organization of Experience* (Cambridge: Harvard Univ. Press, 1974), 10, 347.

6. *Mechanics' Magazine*, Sept. 25, 1830.

7. As discussed in chapter 1, the commercial appearance of artists' prints often reflected the fact that chromolithographs of railroads issued by firms including America's Currier and Ives were sometimes either sold as separate prints or used for commercial advertisements. The idealized views of comfort and safety also mirrored social needs and represented a type of cultural

wish or fantasy—a validation of the century's commitment to social and technological progress.

8. Erasmus Darwin, *The Botanic Garden* (London: J. Johnson, 1791), 1:27.

9. John Dyer, *The Fleece: A Poem in Four Books* (London: R. and J. Dodsley, 1757), 3:101.

10. Ibid., 99. Paul Mantoux in *Industrial Revolution*, 214–215, questioned Dyer's own footnote to his poem attributing the spinning machine to Lewis Paul. Mantoux suggests the inventor may actually have been John Wyatt, who joined forces with Paul. Also, the machine that Dyer saw may have been a cotton- rather than wool-spinning machine, at England's first cotton-spinning mill at Northampton.

11. Dyer, *The Fleece* 1:4.

12. Alexander Pope, *Poetical Works*, ed. Herbert Davis (London: Oxford Univ. Press, 1966), 249.

13. The vision of a rationally designed mechanistic universe was later embellished by Leibniz and satirized by Voltaire. See E. J. Dijksterhuis, *The Mechanization of the World Picture*, trans. C. Dikshoon (London: Oxford Univ. Press, 1969), 490–491. For a discussion of the clock as metaphor, see Samuel L. Macey, *Clocks and the Cosmos: Time in Western Life and Thought* (Hamden, Conn.: Archon, 1980).

14. Arthur Young, *Annals of Agriculture* (1785), vol. 4.

15. "Coalbrookdale 1801, a contemporary description," author unknown. The account appears in the Labouchere Collection of the Shropshire Country Record Office (Ironbridge: Ironbridge Gorge Museum Trust).

16. It was perhaps owing to Williams's gentle portrayal of industry that the artist was commissioned by Abraham Darby, the owner of the ironworks, to create in 1780 commercial drawings of the famed Iron Bridge near Coalbrookdale, the world's first cast-iron bridge, which spanned the River Severn. See Barrie Trinder's introductory essay in Stuart Smith, *A View from the Iron Bridge* (Ironbridge: Ironbridge Gorge Museum Trust, 1979).

17. The explosion, reportedly caused by the overflowing of a pool from a violent thunderstorm, was documented in a handwritten note dating from October 1827 in the Labouchere Collection in the Shropshire Country Record Office.

18. *James Nasmyth, Engineer: An Autobiography*, ed. Samuel Smiles (London: John Murray, 1883), 163.

19. William Wordsworth, *The Excursion*, book 8 (1814). In later poems, Wordsworth took a more conciliatory view of industry.

20. Among the most important studies addressing the absorption, however problematic, of technological images into the natural landscape is Leo Marx, *The Machine in the Garden: Technology and the Pastoral Ideal in America* (New York: Oxford Univ. Press, 1964). See also Marx's essay "The Railroad-in-the-Landscape: An Iconological Reading of a Theme in American Art," in Leo Marx and Susan Danly, *The Railroad in American Art: Representations of Technological Change* (Cambridge, Mass.: MIT Press, 1988); and *The Railroad in the American Landscape: 1850–1950*, exhibition catalog, Susan Danly

Walther, curator (Wellesley, Mass.: Wellesley College Museum, 1981). Another useful discussion of American attitudes toward technology is John F. Kasson, *Civilizing the Machine: Technology and Republican Values in America, 1776–1900* (New York: Grossman, 1976). Barbara Novak's studies of American art present cogent analyses of the presence of technology in American landscape painting. Francis D. Klingender's seminal study *Art and the Industrial Revolution* (1947; rev. and extended by Arthur Elton, 1968; reprint, New York: Schocken Books, 1970) is the best-known discussion of British images of technology in the natural landscape.

21. Earlier antimachine riots occurred after the spinning jenny and other spinning inventions came into use in England. Their unpopularity resulted in riots such as the ones at Leeds in 1780 by workers fearing a lowering of wages. Other riots occurred at Lancashire in 1779 and neighboring areas, where engines were smashed. Mantoux, *Industrial Revolution*, 263–264, 402–403.

22. Klingender, *Art and the Industrial Revolution*. Klingender's study, which focuses on British art, includes a discussion of aesthetic theories of the picturesque and sublime. As formulated by the Rev. William Gilpin and later Sir Uvedale Price in *An Essay on the Picturesque* (1796), the picturesque style typically included landscape images of rough and ragged terrain with battered and weathered ruins in various states of decay. Paintings such as de Loutherbourg's *Coalbrookdale by Night* exemplified in part Edmund Burke's theories of the sublime as presented in *A Philosophical Enquiry into the Origin of Our Ideas of the Sublime and Beautiful*, ed. J. T. Boulton (1757; reprint, London: Routledge and Paul, 1958). Burke saw evidence of the sublime in art that evoked delight as well as terror through fearsome images of the monumental, often cast in an ominous, gloomy darkness.

23. This and other accounts appear in Barrie Trinder, ed., *'The Most Extraordinary District in the World': Ironbridge and Coalbrookdale* (London: Phillimore, 1977), 65, an anthology of visitors' descriptions.

24. Francis Klingender in *Art and the Industrial Revolution*, 98–99, cites contemporary sources in his discussion of de Loutherbourg's eidophusikon. In the introduction to his collection of readings on the British industrial revolution, Humphrey Jennings equates the building of Pandemonium in Milton's poem with "the industrial revolution and the coming of the machine." Jennings, *Pandaemonium 1660–1886: The Coming of the Machine as Seen by Contemporary Observers* (London: Andre Deutsch, 1985), 5. In Milton's poem, the structure of Pandemonium built through the efforts of the metalworkers is modeled after a Greek temple: "where *Pilasters* round / Were set, and Doric pillars overlaid / With Golden Architrave"—an echo of the nineteenth century's coupling of classical design imagery and the machine.

25. American factory engines were slow to supplant readily available waterpower in industries such as textile manufacture, and while there were ironworks producing pig iron and iron bars in colonial America, where ample supplies of wood provided charcoal for smelting, the British iron industry with its use of coke for smelting and lower prices continued to provide substantial imports to the United States. However, although there was

a delayed transfer of iron production technologies to America, by 1815 Pittsburgh began using coke for the reduction of ore, puddling, and rolling. For several useful studies of eighteenth-century and nineteenth-century American industrialization, see Louis C. Hunter, *A History of Industrial Power in the United States, 1780–1930*, vol. 2, *Steam Power* (Charlottesville, Va.: Elutherian Mills/Hagley Foundation, Univ. of Virginia Press, 1985), and Brooke Hindle, *Technology in Early America* (Chapel Hill, N.C.: Univ. of North Carolina Press, 1966). A useful exhibition catalog is Brooke Hindle and Steven Lubar, *Engines of Change: The American Industrial Revolution 1790–1860* (Washington, D.C.: Smithsonian Institution, 1986).

26. W. T. Russell Smith was another artist who recorded new industries, including somewhat romanticized views of the Pittsburgh steel industry. See Virginia E. Lewis, *Russell Smith: Romantic Realist* (Pittsburgh: Univ. of Pittsburgh Press, 1956); for one of the few discussions of this subject, see Marianne Doezema's exhibition catalog essay in *American Realism and the Industrial Age* (Cleveland: Cleveland Museum of Art, 1980), 23–24.

27. Perry Miller, *The Life of the Mind in America from the Revolution to the Civil War* (New York: Harcourt Brace & World, 1965), 269–313. As historians have noted, the United States, with a shortage of skilled labor but ample supplies of wood, water, and coal, was particularly receptive to rapid mechanization.

28. The country's growth was enhanced, too, by the minimal number of restrictive legal and economic precedents as well as by commercial systems of incorporation conducive to business growth. The national program for railroad development was also aided by the lack of restrictive customs barriers and entrenched monopolies. See George R. Taylor, "The Transportation Revolution: 1815–1860," in *The Economic History of the United States* (New York: Rinehart, 1951), 4:74–75; also Douglas C. North, *Economic Growth of the United States, 1790–1860* (Englewood Cliffs, N.J.: Prentice Hall, 1961).

29. British satires of the railroad also appeared in *Punch* after midcentury, and Honoré Daumier in the 1840s published his railroad satires in *Le Charivari*. Several important American comic weeklies, including *Wild Oats* and *Puck*, however, did not begin publication until the 1870s. Discussions of social satire in American art appear in Ralph E. Shikes, *The Indignant Eye: The Artist as Social Critic in Prints and Drawings From the Fifteenth Century to Picasso* (Boston: Beacon Press, 1969).

30. Railroad images did appear in American political cartoons as an emblem of implacable power and an unstoppable machine. In a Currier and Ives cartoon satirizing the election of 1868, Ulysses S. Grant and Schuyler Colfax ride in a locomotive about to run over their opponents Horatio Seymour and Francis Blair. "Clear the track," demands Grant, while Colfax warns, "Take to your heels, neither brass, brag, nor bullets can stop this train."

31. There were other reasons, too, for the dearth of hellish imagery in American artists' views of industry. Although American writers engaged in what Leo Marx has called the "rhetorical technological sublime," the stylistic

conventions of the aesthetic sublime which associated industry with an inferno had largely waned by the 1830s. Later in the century, however, American artists such as Jasper F. Cropsey in his painting *Foundry* (1879) presented more somber views of smoky, fiery industry. Studies of American landscape painting and industry have been scarce. See the exhibition catalog *In Search of the Picturesque: Nineteenth-Century Images of Industry Along the Hudson River Valley* (Annandale-on-Hudson, N.Y.: Edith C. Blum Art Institute, Bard College, 1983). In his catalog essay for this exhibition, Kenneth Maddox noted that a surprising number of American paintings showing industrial scenes were exhibited at the National Gallery of Design before the Civil War but few during the period 1861–1900; he includes a partial list of those depicting furnaces and mills (pp. 19, 29).

32. Ralph Waldo Emerson, "The Poet," in *Essays and Lectures* (1844; reprint, New York: Library of America, 1983), 455.

33. Barbara Novak discusses this balancing of images in *Nature and Culture: American Landscape Painting 1825–1875* (New York: Oxford Univ. Press, 1980), while Leo Marx in *The Machine in the Garden* discusses the often problematic nature of these juxtapositions in American literature.

34. Adams, *Notes on Railroad Accidents*, 232.

35. *Harper's Weekly*, Nov. 18, 1865, 728.

36. John H. White, Jr., *The American Railroad Passenger Car* (Baltimore: Johns Hopkins Press, 1978), 551–553. After the passage of the Railroad Safety Appliance Act in 1893, all American trains were equipped with air brakes.

37. Wolfgang Schivelbush, *The Railway Journey: The Industrialization of Time and Space in the 19th Century*, rev. ed. (Berkeley and Los Angeles: Univ. of California Press, 1986).

38. See Jean Baudrillard, *Simulations*, trans. Paul Foss et al. (New York: Semiotext[e], Inc./Columbia Univ., 1983). See also Baudrillard, *For a Critique of the Political Economy of the Sign*, Charles Levin, trans. (St. Louis: Telos Press, 1981). For further discussion, see Kate Linker, "From Imitation to the Copy to the Just Effect," *Artforum* 22 (Apr. 1984): 44–47.

39. Seymour's robotic gentleman and the depersonalized members of an industrialized society in Charles Dickens's novel *Hard Times* had no major counterpart in nineteenth-century art. In the next century, Charlie Chaplin's film *Modern Times* was a sympathetically comic view of a hapless factory worker's struggles to become machinelike enough to keep up with the demands of an assembly line.

40. For a comprehensive study of American chromolithography, see Peter C. Marzio, *The Democratic Art: Pictures for a 19th-Century America, Chromolithography 1840–1900* (Boston: David R. Godine, 1979).

41. Daguerre's photographic process created a unique image on a silver-plated copper sheet; there was no means to produce multiple copies. Walter Benjamin in "The Work of Art in the Age of Mechanical Reproduction" noted that nineteenth-century observers did not raise "the primary question—whether the very invention of photography had not transformed the entire nature of art." Aaron Scharf in *Art and Photography* discusses the ways

nineteenth-century art was considerably altered by the invention of photography. The Benjamin essay is in Walter Benjamin, *Illuminations*, ed. and intro. by Hannah Arendt, trans. Harry Zahn (New York: Harcourt Brace, 1978, 229); Aaron Scharf, *Art and Photography* (1968; reprint, Penguin, 1986), 13, 128. For a discussion of the role of photography in relation to reproductions, see Miles Orvell, *The Real Thing: Imitation and Authenticity in American Culture 1880–1940* (Chapel Hill: Univ. of North Carolina Press, 1989).

42. John Ruskin, "Modern Manufacture and Design" (1859), in *The Two Paths*, in *The Complete Works of John Ruskin*, ed. E. T. Cook and Alexander Wedderburn (London: George Allen, 1903–1908), 16:340. All further references to Ruskin's work will be to this edition of his works. In the same lecture, Ruskin also suggested that manufacturers could improve design by not pandering to the needs of the market with its appetite for "singularities, novelties, and gaudiness."

43. The critics' complaints about the replacement of the original by the technologically produced simulation were uncannily similar to late twentieth-century concerns about the postmodern predicament—the problematic condition in which simulated, facsimile images were increasingly taking precedence over the handcrafted or even manufactured original. For further discussions, see works by Baudrillard and Jean-Francois Lyotard's exhibition catalog for *Les Materiaux* (Paris: Centre Georges Pompidou, 1985).

44. "Hall Chairs in Cast-Iron Manufactured by Coalbrookdale Iron Company," *Journal of Design and Manufactures* 2 (Sept. 1849–Feb. 1850): 201–202.

45. William Morris, *The Collected Works*, ed. May Morris (London, New York: Longmans, Green, 1910–1915), 90, 87, 93.

46. Charles Dickens, *Our Mutual Friend* (1865), in *The New Oxford Illustrated Dickens* (London: Oxford Univ. Press, 1952, 1959, 1963), 10:6.

47. "Shams and Imitations, Especially in Woven Fabrics," *Journal of Design and Manufactures* 4 (Sept. 1, 1850): 8.

48. As Richard Brown has argued, the experience of the American Revolution had caused new social and material expectations to develop in the United States. The concept of "bettering" oneself had long been part of American colonial society. It was this belief in the legitimacy of social aspirations, along with the "democratization of elite norms," that prompted an embrace of technological modernization in America. See Richard D. Brown, *Modernization: The Transformation of American Life 1600–1865* (New York: Hill and Wang, 1976).

49. William Morris, "Technical Instruction" (1882), in May Morris, *William Morris: Artist, Writer, Socialist* (New York: Russell and Russell, 1966), 1:208–209, 221.

50. John Seegman's study of Victorian Britain, *Consort of Taste, 1830–1870* (1950), attributes the century's extravagant excesses in ornament to Britain's materialism at midcentury and to the middle-class wish to display prosperity. *Consort* has been published under the title *Victorian Taste: A Study in the Arts and Architecture from 1830–1870* (Cambridge, Mass.: MIT Press, 1971), 308.

51. John A. Kouwenhoven, *Made in America: The Arts in Modern Civilization* (1948; reprint, *The Arts in Modern American Civilization*, New York: W. W. Norton, 1967).

52. Kasson, *Civilizing the Machine*, 158–160. In addition to these arguments, Marvin Fisher suggests that American machine ornamentation reflected an underlying conflict in cultural values: industrialism in nineteenth-century America was viewed as a source of social progress and moral improvement, but machines were also feared as a destructive force. This "helps explain why so many nineteenth-century machines and products look so unlike what they were." Fisher, "The Iconology of Industrialism: 1830–60," *American Quarterly* 13 (Fall 1961): 348. See also Fisher, *Workshops in the Wilderness: The European Response to American Industrialization, 1830–1860* (New York: Oxford Univ. Press, 1967).

53. Charles Eastlake, *Hints on Household Taste*, 2d rev. ed. (London: Longmans, Green, 1869), 256, 258.

54. Matthew Digby Wyatt, "Iron-Work and the Principles of Its Treatment," in *Journal of Design and Manufactures* 4 (Sept. 1850–Feb. 1851): 10–11. The article was reprinted in his study *Metal-Work* (London, 1852).

55. It was also argued, as discussed in chapter 3, that improved art education for the general public would also help improve aesthetic taste. There was no point in urging manufacturers to improve the quality of their designs if there was no market of consumers who would appreciate and purchase more elegant designs.

56. Christopher Dresser, "Hindrances to the Progress of Applied Art," *Journal of the Society of Arts* (London), Apr. 12, 1872, 440.

57. "American Art in Tools," *Scientific American*, Oct. 22, 1881, 256.

58. Dresser, "Hindrances," 435.

59. "Artistic Engineering," *Engineering News* 4 (1877): 135.

60. Frank Lloyd Wright, "The Art and Craft of the Machine" (1930), in Lewis Mumford, *Roots of Contemporary American Architecture* (1952; reprint, New York: Dover Books, 1972), 173. All further page references are to the reprint edition.

61. Ibid., 181, 179.

62. Ibid., 179, 181.

1. THE TRAUMAS OF TRANSPORT

1. *Times* (London), July 8, 1808, discussed in H. [Henry] W. [Winram] Dickinson and Arthur Titley, *Richard Trevithick: The Engineer and the Man* (Cambridge: Cambridge Univ. Press, 1934), 106, where the authors state that the exact date and place of the trials cannot be determined. For more on the engine, see pp. 111–113.

2. Frances Ann Kemble, *Records of a Girlhood* (London, 1878; reprint, New York: Henry Holt, 1879), 283, 279.

3. Letter, Sept. 13, 1839, quoted in James Anthony Froude, *Thomas Carlyle: A History of His Life in London, 1834–1881* (London: Longmans, Green, 1884), 1:167.

4. Quoted in Alexander Gilchrist, *The Life of William Blake* (London: MacMillan & Co., 1863).

5. James Renwick, *Treatise on the Steam Engine* (1830), rev. ed. (New York: George Carvill, 1848), 274.

6. "Report of the Special Committee Relative to the Catastrophe in Hague Street, Feb. 4, 1850" (New York: Common Council, City of New York, 1850); Robert Armstrong, *The Modern Practice of Boiler Engineering* (London: E. and F. N. Spon, 1856), note A, 162–167.

7. Report of the Committee of the Franklin Institute, in Armstrong, *Modern Practice*, appendix 2, 187–194.

8. Commissioner of Patents Report, Dec. 30, 1848, as discussed in John G. Burke, "Bursting Boilers and the Federal Power," *Technology and Culture* 7 (Winter 1966): 6–7.

9. U.S. Congress, Senate Committee on Commerce, "Documents relating to the preservation and protection of passengers from injuries resulting from steamboat accidents [1849]," Report by Thibodeaux, Committee of Commerce, 1848, 73.

10. The *Charleston Mercury* report appeared on June 18, 1831, and included a separate report by Thomas Dotterer, an investigating steam engineer. See Samuel Melanchthon Derrick, *Centennial History of the South Carolina Railroad* (Columbia, S.C.: The State Co., 1930), 83–84, 215; C. H. Hewison, *Locomotive Boiler Explosions* (London: David and Charles, 1983).

11. "Steam Carriage," *London Observer*, Sept. 10, 1827.

12. See Gareth Rees, *Early Railway Prints: British Railways from 1825 to 1850* (Oxford: Phaidon, 1980/Ithaca, N.Y.: Cornell Univ. Press, 1980).

13. Marx, *The Machine in the Garden*, 208.

14. There is evidence that Inness idealized the degree of integration, suggested by comparisons of his painting with contemporary photographs of the Scranton landscape, which show a much bleaker vision. See *Railroad in the American Landscape*, 79, and also Nicolai Cikovsky, Jr., "George Inness and the Hudson River School: *The Lackawanna Valley*," *American Art Journal* 2 (Fall 1970), figs. 8–9.

15. In England, the popular colored etchings and engravings satirizing social and political issues had largely disappeared by 1832, the year of the British Reform Bill that brought to a head many of the heated debates that had been a source of graphic satire, but lithographs continued to be an important medium for satire. The most comprehensive discussions of English political and social satire can be found in The British Museum, *Catalogue of Prints and Drawings in the British Museum*, division I: *Political and Personal Satires*, ed. and intro. by M. Dorothy George. See in particular vol. 11, 1828–1832 (London, 1954). See also M. Dorothy George, *Hogarth to Cruikshank: Social Change in Graphic Satire* (New York: Walker and Co., 1967), 17, and *English Political Caricature*, vol. 2 [1793–1832] (Oxford: Oxford Univ. Press, 1959). An overview of the subject appears in the exhibition catalog *English Caricature 1620 to the Present* (London: Victoria and Albert Museum, 1984).

16. Darwin, *Botanic Garden* 1:29.

17. British Museum, *Catalogue* 9:liv and 11:xliv–xlv; George, *English Political Caricatures* 2:230.

18. "Aeropleustics, or Navigation in the Air," *London Observer*, Sept. 30, 1827.

19. "Mr. Gurney's New Steam Carriage," *London Observer*, Dec. 9, 1827.

20. J. Willoughby Gordon, "Memorandum" (1829), in Sir Herbert Taylor, *The Taylor Papers* (London: Longmans, Green, 1913), 263.

21. Lt. Colonel Sir Charles Dance, "Mr. Gurney's Steam Carriage on the Bath Road," in *Taylor Papers*, 267. In an 1831 report, an investigation by Britain's House of Commons declared the steam carriage to be safer than horse-drawn carriages and pointed to the development of a "sectional" boiler, which was said to ensure against injury by explosion. See Robert H. Thurston, *A History of the Growth of the Steam-Engine* (1878), centennial ed. (Ithaca, N.Y.: Cornell Univ. Press, 1939), 171.

22. See Francis T. Evans, "Roads, Railways, and Canals," *Technology and Culture* 22 (Jan. 1981): 26–32. Although twentieth-century British accounts of the steam coach generally attribute its demise to social and economic causes, Evans concludes that "it was technology rather than social or economic institutions which stymied steam coaches." He notes that the success of White steam cars and Stanley steam cars built at the end of the nineteenth century was due to advances in machine tools, steel boilers, and the use of liquid fuels (p. 32).

23. Biographical information on Seymour is scant, but a brief useful sketch of the artist is presented in F. Gordon Roe, "Portrait Painter to 'Pickwick'; or Robert Seymour's Career," *Connoisseur* 77 (Mar. 1927): 152–157. The controversies over Seymour's illustrations for *The Pickwick Papers* and the grievances that troubled his last years were discussed in his widow Jane Seymour's pamphlet, "An Account of the Origin of the 'Pickwick Papers'" (reprint, London: F. G. Kitton, 1901).

24. Frederick Marryat, *A Diary in America* (London, 1839; reprint, ed. Jules Zanger, Bloomington: Indiana Univ. Press, 1960), 101.

25. Samuel Breck, in *Recollections of Samuel Breck with Passages from His Note-Books 1771–1862*, ed. H. E. Scudder (Philadelphia: Porter and Coates, 1877), 275.

26. Ibid., 277.

27. John B. Jervis, *Railway and Locomotive Historical Society Bulletin* no. 55 (May 1941), 10, quoted in John H. White, Jr., *American Locomotives: An Engineering History 1830–1880* (Baltimore: Johns Hopkins Press, 1968), 73.

28. Ibid., 73–74. White notes that the New York averages were probably above the national average.

29. For further discussion on psychic traumas, see Schivelbusch, *The Railway Journey*.

30. Henry David Thoreau, *Walden* (1854; reprint, Boston: Houghton, 1898) 180, 185–186.

31. Charles Dickens, *American Notes for General Circulation* (1842), in *New Oxford Illustrated Dickens* (London: Oxford Univ. Press, 1952, 1959, 1963), 19:64.

32. Charles Dickens, "No. 1 Branch Line. The Signalman," in "All the Year Round" (1863; reprint as the second of "Two Ghost Stories," in *Christmas Stories from "Household Words" and "All the Year Round*," London: Chapman and Hall, 1879). In his talk "Transport in the Dickensian Era," T. S. Lascelles suggests that "The Signalman" referred to an accident in the Clayton Tunnel near Brighton in 1861. *The Dickensian* 58 (May 1962): 75–86 and part 2 (Sept. 1962): 152–160 (the reference is to p. 157).

33. The engine was later rebuilt, with modifications. Reports quoted in Derrick, *Centennial History*, 83–84; L. T. C. Rolt, *Red For Danger: A History of Railway Accidents and Railway Safety Precautions* (London: Bodley Head, 1955), 55, 59–60.

34. Dionysius Lardner, ed., *The Museum of Science and Art* (London: Walton and Maberly, 1854–1856), 181–182, 184, 189.

35. *Harper's Weekly*, Feb. 10, 1872. The illustrator is noted as W. L. Sheppard.

36. Bernard F. Reilly, Jr., introductory essay to *Currier and Ives: A Catalogue Raisonné* (Detroit: Gale Research Co., 1984), 1:25. *An American Railway Scene at Hornellsville* was by artists Charles R. Parsons and Lyman W. Atwater and was issued in 1874 as a stock print without advertising and in 1876 with advertising. See also Harry T. Peters, *Currier and Ives, Printmakers to the American People* (Garden City, N.Y.: Doubleday, Doran, 1942).

37. Marzio, *The Democratic Art*, 192–193.

38. *Harper's Weekly*, Sept. 20, 1873.

39. The "Darktown" railroad comics were illustrated in Fred J. Peters, comp., *Railroad, Indian and Pioneer Prints by Nathan Currier and Currier and Ives* (New York: Antique Bulletin, 1930).

40. John F. Stover, *American Railroads* (Chicago: Univ. of Chicago Press, 1961), 2; Henry Nash and Smith, *Virgin Land: The American West as Symbol and Myth* (Cambridge, Mass.: Harvard Univ. Press, 1950). For a cogent study of the railroad's impact on American culture, see John Stilgoe, *Metropolitan Corridor: Railroads and the American Scene* (New Haven, Conn.: Yale Univ. Press, 1983).

41. John Stover noted that a similar trauma and feeling of national shock was experienced in the United States when the Pacific Express on Vanderbilt's Lake Shore & Michigan Southern crashed through a bridge in 1876 at Ashtabula, Ohio—causing dozens to die in the ensuing fire. *American Railroads*, 168–169.

42. Useful studies of Daumier's work include Roger Passeron, *Daumier* (New York: Rizzoli, 1981); Oliver W. Larkin, *Daumier: Man of His Time* (Boston: Beacon Press, 1966); and Howard P. Vincent, *Daumier and His World* (Evanston, Ill.: Northwestern Univ. Press, 1968).

43. Charles Baudelaire, "Some French Caricaturists," in Jonathan Mayne, trans. and ed., *The Painter of Modern Life and Other Essays* (1964; reprint, New York: Garland, 1978), 188.

44. John H. White, Jr., *The American Railroad Passenger Car* (Baltimore: Johns Hopkins Univ. Press, 1978), 203. White quotes a *Scientific American* article of October 2, 1852, which insisted, "we have no second class cars, for

the inferior classes, because all our citizens rank as gentlemen." Black passengers, however, were often required to sit in the baggage car.

45. Ruskin, *Complete Works* 8:159.

46. White, *American Railroad Passenger Car*, 218–251. As White notes, more comfortable coach seats were widely used in America by 1885. Increased long-distance train trips brought a renewed American interest in previously introduced sleeping cars, including Theodore Woodruff's "palace car" sleeper, first in use in 1864, soon followed by George Pullman's better-known "Pioneer." Woodruff's "Silver Palace" car of 1866 had black walnut interiors, silverplated lamps and hardware, and velvet wall hangings. Luxury parlor cars were similarly richly appointed with wood-paneled walls, ornamentally painted ceilings, and plush velvet seats. Dining cars were introduced in the 1860s on many American trains in the Northeast and Great Lakes states. For a cogent early discussion of patent railroad furniture, see Siegfried Giedion, *Mechanization Takes Command* (New York: Oxford Univ. Press, 1949), 440–467.

47. Hamilton Ellis, *Railway Carriages in the British Isles: From 1830 to 1914* (London: George Allen and Unwin, 1965), 57–58. The British bought their first sleeping cars from Pullman in 1874, but as John H. White observed, the cars were not well received due to the relatively short runs of British rail travel. *American Railroad Passenger Car*, 226.

48. After introducing third-class upholstered seating in 1874, England's Midland Railroad Company also took the added steps of merging second and third class, with the second class disappearing in 1875, and raising third-class standards to those of second class, a practice grudgingly followed by other British railroads. Ellis, *Railway Carriages*, 65–66. Ellis notes that the Midland's manager, Sir James Allport, was ridiculed by Tory critics who saw his innovations as a "blow at the sanctity of caste and a very bad business step, likely to lose the Midland company its bourgeois user" (p. 65).

49. Sir John Tenniel's woodcut "Death and His Brother Sleep" (*Punch*, Oct. 4, 1890) depicts a skeleton taking over the controls of a speeding train—a reference to Shelley's poem and to a railway collision in 1890 at Eastleigh, discussed in the exhibition catalog *Train Spotting: Images of the Railway in Art* (Nottingham: Nottingham Castle Museum, 1985), 38. The association of a macabre skeleton and a speeding railroad, a recurring motif in nineteenth-century art, also appeared in Daumier's very rare lithograph "Madame Deménage!" (1867), where a sickle-carrying skeleton straddles a locomotive. Loys Delteil noted that publication of the lithograph was censored. Delteil, *Le Peintre-Graveur Illustré*, illus. no. 3590 (Paris: Delteil, 1926). One of the best-known British paintings presenting warmth and intimacy in a station interior, W. P. Frith's *The Railway Station* (1863), encompasses the huge interior of London's Paddington Station with its tight crowds of passengers caught in melodramatic poses. In the center foreground, a mother dressed in a red shawl bends to kiss her son farewell in a moment of focused intimacy beneath the massive vaulted ceiling of the station.

50. Solomon's second version of the flirting couple is titled *First Class: The Meeting . . . And at First Meeting Loved* (1854). His painting *Second Class:*

The Parting . . . Thus Part We Rich in Sorrow, Parting Poor (1855) presents two central figures of a mother with her arm around her sad-faced son. These and other British Victorian scenes of railroad interiors and railroad caricatures are illustrated in *Train Spotting*. See also C. Hamilton Ellis, *Railway Art* (Boston: New York Graphic Society, 1977), for a popular overview of the railroad in art.

51. The theme of railroad carriages as a place for illicit intimacies also found its way into American artists' illustrations. The cover of *Frank Leslie's Weekly* on January 20, 1898, featured a spacious, ornate parlor car in which a blushing woman traveler sits between two seated gentlemen as the coy caption asks, "Who kissed Her in the Tunnel?"

2. ART, TECHNOLOGY, AND THE HUMAN IMAGE

1. *The Crystal Palace and Its Contents: An Illustrated Cyclopedia of the Great Exhibition of Industry of All Nations, 1851* (London: W. M. Clark, 1852), 87.

2. *Great Exhibition of the Works of Industry of All Nations, 1851, Reports by the Juries* (London: Wm. Clowes, 1852), 1:684–685.

3. *Official Catalogue of the Great Exhibition of the Works of Industry of All Nations I* (London: Spicer, 1851), 433.

4. Lewis Mumford, *Technics and Civilization* (1934; reprint, New York: Harcourt Brace, 1963), 162.

5. William Dean Howells, discussed and quoted in Alan Trachtenberg, *The Incorporation of America: Culture and Society in the Gilded Age* (New York: Hill and Wang, 1982), 47.

6. Charles T. Porter, *Engineering Reminiscences* (New York: John Wiley, 1908), 248–250.

7. *Harper's Weekly*, Feb. 23, 1889, 152–153. Graham's illustration "Making Bessemer Steel at Pittsburgh" appeared in its uncut form in *Harper's Weekly*, Apr. 10, 1886, 232–233.

8. Archives of American Art, roll 529, frames 933–938.

9. Thomas Carlyle, "Signs of the Times," *Edinburgh Review* 49 (June 1829): 444. Also in *Critical and Miscellaneous Essays* (Boston: J. Munroe, 1839), 2:143–171.

10. René Descartes, *The Philosophical Writings of Descartes*, 2 vols., trans. E. S. Haldane and G. R. T. Ross (Cambridge: Cambridge Univ. Press, 1967); Descartes, *Discourse on Method*, trans. F. E. Sutcliffe (Harmondsworth: Penguin, 1968), 73–74; Descartes, *Philosophical Letters*, trans. Anthony Kenny (Oxford: Clarendon Press, 1970), 53–54.

11. Julien Offray de La Mettrie, *L'Homme machine* (1748) (Man a machine), trans. Gertrude Carman Bussey (La Salle, Ill.: Open Court Pub., 1912, 1961), 135, 93.

12. Giovanni Branca, *Le Machine: volume nuovo et di multo artificio da fare effetti maravigliosi, tanto spiritali quanto di animale operatione . . .*, illus. by Monsignor Tiberio (Rome: J. Manuci, 1629), fig. 25.

13. Homer, *The Iliad*, prose trans. E. V. Rieu (Harmondsworth/New York: Penguin Books), 348.

14. In *L'Homme machine*, La Mettrie wrote admiringly of Vaucanson's au-

tomatons and saw in them an affirmation of his own mechanistic theories, likening the complexity that would be needed to create a talking automaton (which he considered technically feasible) as comparable to the skill and complexity used by nature to create the mechanism of a human being: "In like fashion, it was necessary that nature should use more elaborate art in making and sustaining a machine which for a whole century could mark all motions of the heart and of the mind" (pp. 140–141).

15. Alfred Chapuis and Edmond Droz, *Automata: A Historical and Technological Study*, trans. Alec Reid (Neuchâtel/London: B. T. Batsford, 1958). The poem is quoted on p. 284. For additional studies of automatons, see Alfred Chapuis and Edouard Gelis, *Le Mondes des automates* (Paris, 1928; reprint, Geneva-Paris: Editions Slatkine, 1984); John Cohen, *Human Robots in Myth and Science* (London: George Allen and Unwin, 1966); and Marvin Minsky, ed., *Robotics* (Garden City, N.Y.: Anchor Press/Doubleday, 1985).

16. See John A. Kouwenhoven, "Who's Afraid of the Machine in the Garden?" in *Half a Truth Is Better Than None* (Chicago: Univ. of Chicago Press, 1982), 136–139. Thomas D. Clareson in his bibliography *Science Fiction in America 1870s–1930s: An Annotated Bibliography of Primary Sources* (Westport, Conn.: Greenwood Press, 1984), 92, lists Ellis's novel as originally published as one of *Irwin's American Novels* (1865); Kouwenhoven dates it as August 1868. An 1882 edition of Ellis's novel was titled *The Huge Hunter; or, The Steam Man of the Prairies* (New York: Beadle and Adams, 1882). Kouwenhoven lists Lu Sanarens as possible author of *Frank Reade and His Steam Man of the Plains* (1883), the first in a new series of steam man dime novels.

17. See Blair Whitton, *Clockwork Toys 1862–1900* (Exton, Pa.: Schiffler, 1981), 21.

18. Ibid., 17. Whitton notes that beginning in the 1850s, German manufacturers began producing less expensive spring-wound dolls made of machine-stamped tin parts (p. 209). It was these versions which eventually undercut American manufacturers of clock-driven toys.

19. See David Hounshell, *From the American System to Mass Production, 1800–1932: The Development of Manufacturing Technology in the United States* (Baltimore: Johns Hopkins Univ. Press, 1984); Otto Mayr and Robert C. Post, *Yankee Enterprise: The Rise of the American System of Manufactures* (Washington, D.C.: Smithsonian Institution Press, 1981); and Donald R. Hoke, *Ingenius Yankees: The Rise of the American System of Manufactures in the Private Sector* (New York: Columbia Univ. Press, 1990).

20. Whitton reports that probably about one hundred of the dolls were manufactured between 1880 and 1890. *Clockwork Toys*, 180.

21. "Edison's Phonographic Doll," *Scientific American*, Apr. 26, 1890.

22. See Chayim Bloch, *The Golem: Mystical Tales from the Ghetto of Prague* (Blauvelt, N.Y., 1972), and Mosche Idel, *Golem: Jewish Magical and Mystical Traditions on the Artificial Anthropoid* (Albany: State Univ. of New York Press, 1990).

23. For essays on scientific sources in *Frankenstein*, see Samuel Holmes Vasbinder, *Scientific Attitudes in Mary Shelley's "Frankenstein"* (Ann Arbor, Mich.: UMI Research Press, 1976, 1984), and James Rieger, introductory essay to

Frankenstein; or, The Modern Prometheus (1818 text), ed. James Rieger (Indianapolis/New York: Bobbs-Merrill, 1974). Diane Johnson, in her introductory essay to *Frankenstein* (New York: Bantam Books, 1981), discusses the impact of Shelley's infants' deaths on her writing.

24. Mary Shelley, *Frankenstein; or, The Modern Prometheus* (1818), rev. ed. 1836, ed. M. K. Joseph (reprint, London, New York: Oxford Univ. Press, 1969), 94–96. All page references are to this edition.

25. Ibid., 98, 99, 204.

26. "The Artificial Man," *Scientific American* 1 (Oct. 8, 1859): 285.

27. Henri Bergson, *Le Rire* (1900; reprint, Wylie Sypher, ed., *Comedy: An Essay*, New York: Doubleday, 1956), 218.

28. Sigmund Freud, *Civilization and Its Discontents*, trans. James Strachey (New York: W. W. Norton, 1961), 38–39.

29. Little has been written about Cham (Amédée de Noé). See Felix Ribeyre, *Cham: Sa vie et son oeuvre* (Paris: E. Plon, Nourrit, 1884), and David Kunzle, "Cham: The 'Popular' Caricaturist," *Gazette des Beaux-Arts* 96 (Dec. 1980): 213–224.

30. Ruskin, *Complete Works* 10:192.

31. Herman Melville, "The Paradise of Bachelors and the Tartarus of Maids" (1855), in *The Writings of Herman Melville*, ed. Harrison Hayford et al. (Evanston and Chicago: Northwestern Univ. Press and Newberry, 1987), 9:316–335. The quote is from p. 328.

3. TECHNOLOGY AND THE DESIGN DEBATE

1. George Eliot, *Felix Holt, The Radical* (1866; reprint, *The Works of George Eliot*, New York: Nottingham Society, n.d.), 5:3–8.

2. "The Mutual Interests of Artists and Manufacturers," *Art-Union* 10 (Mar. 1, 1848): 69.

3. Ruskin, *Complete Works* 34:521.

4. Ibid., 16:340–341. The lecture was presented before the Mechanics' Institute in Bradford on Mar. 1, 1859, for the inauguration of the new School of Design.

5. "Mutual Interests," 70.

6. Ibid., 69.

7. Ibid., 69–70.

8. John Heskett has argued that there were fundamental differences in modes of production between British and American manufacturers of the decorative arts. British manufacturers beginning in the late eighteenth century tended to adapt crafts techniques and traditions to mass production, making changes in the realm of "commercial organization and production, rather than in the manufacturing methods by which goods were produced." At the factories of Boulton and Wedgwood, there was specialization of labor which included skilled craftsmen who put decorative finishes on ornamental wares. American manufacturers, however, were most focused on developing mass production methods based on the American System, or large-scale manufacture of standardized products with interchangeable parts, using powered machine tools in a sequence of simplified

mechanical operations. Heskett, *Industrial Design* (New York and Toronto: Oxford Univ. Press, 1980), 50.

9. Judith Banister in *Old English Silver* (London: Evans Bros., 1965), 25, argues that stamping was not widely used for producing silverware until the middle of the nineteenth century. However, Eric Robertson has written that Matthew Boulton was an exception to this generalization, for there is evidence of his using stamping in the 1770s to manufacture silverplated candlesticks at his Birmingham factory. See Robinson, "Problems in the Mechanization and Organization of the Birmingham Jewelry and Silver Trades, 1760–1800," in *Technological Innovation and the Decorative Arts*, ed. Ian M. G. Quimbly and Polly Anne Earl, Winterthur Conference Report 1973, the Henry Francis du Pont Winterthur Museum (Charlottesville: Univ. Press of Virginia, 1974), 78–80.

10. Robinson, "Problems in Mechanization," 75–76. Robinson cites Charles Holtzapffel, *Turning and Mechanical Manipulation*, 6 vols. (London: Hotlzapffel and Co., 1846–1884), vol. 1. By 1820, the industry was also making use of spinning lathes to shape articles. See Shirley Bury, *Victorian Electroplate* (London: Country Life Books, 1971), 33.

11. Cyril Stanley Smith, "Reflections on Technology and the Decorative Arts in the Nineteenth Century," in Quimbly and Earl, *Technological Innovation*, 4, 7–8.

12. R. A. F. de Réaumur, *L'Art de Convertir le Fer Forgé en Acier, et l'Art d'Adoucir le Fer Fondu ou de Faire des Ouvrages de Fer Fondu Aussi Finis que de Fer Forgé* (Paris: Michel Brunet, 1722; reprint, English trans. Anneliese Grunhaldt Sisco, Chicago: Univ. of Chicago Press, 1956), 352.

13. "Carving by Machinery," *Art-Union* 10 (June 1, 1848): 194. Jordan's remarks, written in a letter to the Society of Arts, appeared in the Society's *Transactions*, 1846–1847, part I, 160. Jordan's church screen produced by his patent carving machine is illustrated in Nikolaus Pevsner, *High Victorian Design: A Study of the Exhibits of 1851* (London: Architectural Press, 1951), 32.

14. "Carving by Machinery," 194.

15. Ruskin, *Complete Works* 8:60, 81, 83.

16. Ibid., 219.

17. Ibid., 6:333.

18. Christopher Dresser, *Principles of Decorative Design* (London/New York: Cassell, Petter, and Galpin, 1873), 138, 136–137. Originally published as a series of lectures in Cassell's *Technical Educator* (London, 1870–1873).

19. Horace Greeley, *Art and Industry as Represented in the Exhibition at the Crystal Palace—1853–54* (New York: Redfield, 1853), 52–53.

20. It can also be argued that because of America's relatively brief history and quest for its own cultural base, the country's critics were less likely to tout medieval handicraft traditions.

21. Smith, "Reflections on Technology," 10–13.

22. Morris, "The Arts and Crafts of To-Day," address delivered in Edinburgh in 1889 before the National Association for the Advancement of Art, in *Collected Works* 23:366.

23. Ibid., 373.

24. Ruskin, *Collected Works* 10:192.
25. Morris, "Arts and Crafts of To-Day," 368.
26. Matthew Digby Wyatt, "Iron-Work, and the Principles of Its Treatment," *Journal of Design and Manufactures* 4 (Sept. 1850–Feb. 1851): 78, and reprinted in Wyatt's *Metal-Work*, xix.
27. "Illustrated Tour in the Manufacturing Districts: The Ironwork of Coalbrookdale," *Art-Union*, Aug. 1, 1846, 220.
28. Wyatt, "Iron-Work," 10–11.
29. "Universal Infidelity in Principles of Design," *Journal of Design and Manufactures* 5 (Mar.–Aug. 1851): 158–160. Nikolaus Pevsner in *High Victorian Design*, 151, assumes that the *Times* article reprinted in the *Journal* was written by a member of Henry Cole's circle, which included Richard Redgrave and Owen Jones, among others; Siegfried Giedion also comments that the article "reflects the circle around Henry Cole" (*Mechanization Takes Command*, 351).
30. "Universal Infidelity," 158. A call for a functional or utilitarian focus in industrial design was addressed in the writings of Gottfried Semper, a German architect living in London during the period of the Great Exhibition, and in the report written by Richard Redgrave, a member of the Cole group of reformers and an editor of the *Journal of Design*. See Richard Redgrave, "Supplementary Report" in *Great Exhibition of the Works of Industry* 2:1547–1682.
31. Walter Smith, "Industrial Art Education," lecture delivered in Philadelphia, Apr. 23, 1875 (Boston: L. Prang and Co., 1875; reprint, *Penn Monthly*, July 1875), 6; Walter Smith, "Industrial Education and Drawing as Its Basis," lecture delivered before the Massachusetts Teachers' Association, Worcester, Mass., Dec. 28, 1878 (Normal Art School, 1878), 31, quoted in Isaac Edwards Clarke, ed., *Art and Industry: Education in the Industrial and Fine Arts in the United States*, 4 vols. (Washington, D.C.: United States Government Printing Office, 1885–1898), 1:xcv.
32. "Art and Industry. The Democracy of Art. Preliminary Papers upon the Relations of Art to Education, Industry, and National Prosperity," quoted in Clarke, ed., *Art and Industry* 1:xciii, xcv.
33. See Heskett, *Industrial Design*, 11.
34. Ibid., 10–12. For added material on encyclopedias of ornament, see Stuart Durant, *Ornament: From the Industrial Revolution to Today* (Woodstock, N.Y.: Overlook Press, 1986).
35. Heskett, *Industrial Design*, 14–15. In their two-part article, "The Matthew Boulton Pattern Books," W. A. Seaby and R. J. Hetherington cite several artists working for Boulton, including Robert and James Adam, John Flaxman, Francis Eginton, and George Wyon. *Apollo* 51 (Jan. 1950): 48–50 and 52 (Mar. 1950): 81.
36. Bury, *Victorian Electroplate*, 42–44.
37. Smith, "Reflections on Technology," 17.
38. "Mutual Interests of Artists and Manufacturers," 69.
39. The tours were first reported in *Art-Union*, in 1846 and 1847, and

were recounted in "Art and Art-Manufacture, 1851–1877," *Art Journal* (London) 39 (1877): 103.

40. Morris, *Collected Works*, 221.
41. Quentin Bell, *The Schools of Design* (London: Routledge and Kegan Paul, 1963), 52.
42. Report of Select Committee of 1835, quoted ibid., 52.
43. Ibid., 46–47.
44. *Journal of Design and Manufactures* 1 (1849): 24–26.
45. Ruskin, *Complete Works* 15:344.
46. U.S. Department of the Interior Bureau of Education, "Education in the Industrial and Fine Arts, the Technical Education of a People," in Clarke, ed., *Art and Industry* 2:xcv, xcvi. Clarke quoted from Walter Smith's talk "Industrial Education and Drawing as Its Basis."
47. Clarke, ed., *Art and Industry*, vol. 3.
48. Walter Smith, *Art Education, Scholastic and Industrial* (Boston: Osgood, 1872), 189.
49. "Contributions to the International Art Exhibition, Philadelphia," *Art Journal* (London) 38 (1876): 341.
50. "The Development of Modern Industrial Art in Germany, The Munich Exhibition," *Art Journal* (London) 51 (1889): 38.
51. Christopher Dresser, "Ornamentation Considered as High Art," *Journal of the Society of Arts* 19 (Feb. 10, 1871): 217.
52. Dresser, "Hindrances," 435–443. Quotations from 435, 440.
53. Dresser, "Art Schools," the last of three lectures sponsored by the Philadelphia Museum and School of Industrial Art, *Penn Monthly*, Mar. 1877, 221, 219.
54. Ibid., 221, 225.
55. Dresser, "Hindrances," 435, 440.
56. Dresser, *Principles of Decorative Design*, 138.

4. THE ANXIETY OF IMITATION

1. *Art-Journal Illustrated Catalogue of the Great Exhibition of Industry of All Nations* (London: G. Virtue, 1851), 195.
2. Walter Smith, "The Industrial Art of the International Exhibition," in *Examples of Household Taste: The Industrial Art of the International Exhibition* (New York: R. Worthington, [1884]), 2:62–64.
3. Greeley, *Art and Industry*, 52.
4. Ruskin, *Complete Works* 8:81, 83.
5. A useful historical overview of metal technologies, including electrometallurgy and the decorative arts, appears in Cyril Stanley Smith's "Reflections on Technology," 1–64. As Smith notes, although Jordan, Jacobi, and Spencer all recognized electrotyping's usefulness, Jacobi was the first to make a public announcement about it in 1838 (pp. 28–31). All three first published their findings in 1839. See Jordan, "Engraving by Galvanism," *Mechanics' Magazine,* June 8, 1839; reprint in James Napier, *Manual of Electro-Metallurgy* (London: Greenwich, 1851); Moritz Hermann von Jacobi

(Boris Semenovich Jacobi), "On the Method of Producing Copies of Engraved Plates by Voltaic Action," *Philosophical Magazine* 15 (1839): 161–165; Thomas Spencer, *An Account of Some Experiments Made for the Purpose of Ascertaining How Far Voltaic Electricity May Be Usefully Applied to the Purpose of Working in Metal* (Liverpool: Mitchell, 1839). For a discussion of Jacobi's work, see Olga I. Pavlova, *Electrodeposition of Metals: A Historical Survey*, ed. S. A. Pogodin (Moscow: Izdatel'stvo Akademii Nauk SSSR, 1963). After sending his report on the invention of electrotyping to the St. Petersburg Academy of Sciences in 1838, Jacobi in 1839 submitted to the academy an electrotyped copper bas-relief of a scene from Homer's *Odyssey* copied from an original sculpture by the artist F. P. Tolstoi as well as an electrotyped copy of a sculpture by Bernini (Pavlova, iv, 27–28).

6. The fascination with large-scale electrotypes waned by the end of the nineteenth century, though electrotypes of small objects were still being produced at later dates.

7. In *Mechanization Takes Command*, Siegfried Giedion cites several British patents for molding, stamping, and embossing, including patents issued in 1838, 1844, and 1846 (pp. 346–347). Cyril S. Smith in "Reflections on Technology" argues that the stamped metals from worn dies "brought stamping to a disrepute that the technique per se does not deserve" (p. 10). For another useful study by Smith, see "Art, Technology, and Science: Notes on Their Historical Interaction," *Technology and Culture* 2 (Oct. 1970): 493–549.

8. Theophilus, *De diversis artibus*, manuscript treatise c. 1125 A.D.; trans. with technical notes by J. G. Hawthorne and C. S. Smith (Chicago: Univ. of Chicago Press, 1963).

9. For extensive studies of silverplating, see Frederick Bradbury, *History of Old Sheffield Plate* (London: Macmillan, 1912), and Patricia Wardle, *Victorian Silver and Silverplate* (London: Barrie and Jenkins, 1963), 29–32ff.

10. Nickel silver, like britannia, had the advantage that when the plating was worn away, the base metal would be less noticeable. Discussions of nineteenth-century electroplating and the decorative arts are scarce. See Smith, "Reflections on Technology," and Bury, *Victorian Electroplate*. An extensive study of both silverplating and electroplating appears in Wardle's *Victorian Silver and Silverplate*.

11. "Messrs. Elkington, Mason & Co.'s Electro-Plate Works," *Illustrated Exhibitor and Magazine of Art* (London) 1 (Jan. 3, 1852): 295–300.

12. Bury, *Victorian Electroplate*, 26–27.

13. *The Furniture Gazette* (London), Feb. 5, 1876, 78.

14. See George Sweet Gibb, *The Whitesmiths of Taunton: A History of Reed & Barton 1824–1943* (Cambridge, Mass.: Harvard Univ. Press, 1943), 128. In 1847, William Rogers had already begun manufacturing electroplated flatware, keeping the firm's trademark even after the company became largely absorbed by Meriden Britannia Co. in 1862.

15. Rogers Bros. was the first to manufacture electroplated flatware in 1847, though in 1862 the firm brought its manufacturing equipment to join Meriden Britannia, which kept using the Rogers trademark for its flatware. Reed & Barton began manufacturing electroplate on britannia hollowware

at the end of the 1840s, and Meriden Britannia was organized in 1853. As a common practice of the period, firms such as Reed & Barton would send their own hollowware forms to Rogers Bros. for plating, and Reed & Barton bought its flatware from Rogers Bros., issuing it under the Reed & Barton trademark. For an overview of American electroplate manufacturers, see Dorothy T. Rainwater and H. Ivan, *American Silverplate* (Nashville: T. Nelson, 1972).

16. "Silver and Silver Plate," *Harper's New Monthly Magazine* 37 (Sept. 1868): 445–448.

17. Charles J. Carpenter, Jr., *Gorham Silver 1831–1891* (New York: Dodd, Mead, 1982), 86; Meriden Britannia Company electroplate catalog of 1867, reproduced in *Victorian Silverplated Holloware* (Princeton, N.J.: Pyne Press, 1972).

18. Gibb, *Whitesmiths*, 131–132.

19. Augustus Welby Pugin, *The True Principles of Pointed or Christian Architecture* (London: John Weale, 1841), 2.

20. Ruskin, *Complete Works* 8:219–220.

21. Ibid., 86.

22. Ibid., 219, 82.

23. Pugin, *True Principles*, 27.

24. Dickens, *Our Mutual Friend* 10:131.

25. *Journal of Design and Manufactures* 5 (Mar. 1, 1851): 54.

26. Ibid., 4 (Sept. 1850–Feb. 1851): 150–51.

27. Eastlake, *Hints on Household Taste*, 255, 254, 264.

28. "The Electrotype," *The Art-Union* 3 (Jan. 1841): 14.

29. *Art-Journal Illustrated Catalogue of the Great Exhibition*, xvi.

30. Greeley, *Art and Industry*, 49–50.

31. Ibid., 50.

32. Ibid.

33. John Leander Bishop, *A History of American Manufactures from 1608 to 1860*, 3d rev. ed. (Philadelphia: Edward Young Co., 1868), 331.

34. Benjamin Silliman, Jr., and Charles Goodrich, eds., *The World of Science, Art, and Industry illustrated from examples in the New York exhibition, 1853–54* (New York: G. P. Putnam, 1854), 9.

35. Greeley, *Art and Industry*, 52–53.

36. *Frank Leslie's Historical Register* (1876): 303. Electroplating did not have hallmarks, but was frequently marked with the letters EP, EPNS (electroplate on nickel silver), EPES.

37. *The Art Journal* (American ed.), 1876, 304.

38. "Silver and Silver Plate," 433–448, 434.

39. Ibid., 434, 448.

40. An illustration of the Gorham broadside appears in Carpenter, *Gorham Silver*, 87. The leading nineteenth-century American electroplate manufacturers regularly sold hollowware to jobbers who then put on their own plating and trademarks—a practice that no doubt contributed to anxieties about counterfeit. See Dorothy Rainwater, *Encyclopedia of Silverplate*, 3d rev. ed. (Westchester, Pa.: Schiffler, 1986), 157.

41. "Silver and Silverplate," 441–442

42. Ibid., 445.

43. Dresser himself provided a design for Jones's *Grammar of Ornament*, which was published as plate 98, presenting an abstracted geometric arrangement of flowers.

44. For discussions of Dresser's electroplate designs, see Nikolaus Pevsner, "Christopher Dresser: Industrial Designer," *Architectural Review* 81 (1937): 183–186; Shirley Bury, "The Silver Designs of Dr. Christopher Dresser," *Apollo* (Dec. 1962); Bury, *Victorian Electroplate*; *Christopher Dresser*, exhibition catalog (London: The Fine Arts Society, 1972); and *Christopher Dresser*, exhibition catalog (Dorman Museum and the Camden Arts Center, England, 1979).

45. *Art Journal* (London) 18 (1879): 222.

5. THE STRUGGLE FOR LEGITIMACY

1. "Exposition of British Industrial Art at Manchester," *The Art-Union* 8 (supplementary number, Jan. 1846): 47.

2. Ruskin, *Complete Works* 8:85.

3. For detailed nineteenth-century accounts of the cast-iron process, see G. W. Yapp, *Art Industry* (London, 1879), and *Iron Castings: Art Metal and Constructional Iron Work: The Works of the Coalbrookdale Company: Reprinted from "The British Mercantile Gazette"* (London: The British Mercantile Gazette, 1878). A more recent account is in E. Graeme Robertson and Joan Robertson, *Cast Iron Decoration: A World Survey* (New York: Whitney Library of Design, 1977).

4. Details of Prussian cast-iron manufacture can be found in Zdenke Rasl, *Decorative Cast Ironwork: Catalogue of Artistic and Decorative Iron Castings from the 16th to 20th Centuries Preserved at the National Technical Museum of Prague* (Prague, 1980); see 30–42. Rasl notes that portrait busts were cast at the German foundry at Lachhammer as early as 1765 and that vases, candlesticks, and other wares were being produced at Berlin and Gliwice (Gleiwitz) by 1800, with production reaching its peak after 1814. See also Gail C. Andrews, "Prussian Artistic Cast Iron," *Antiques* 120 (no. 2, Feb. 1983), 422–427.

5. *A New Phase in the Iron Manufacture*, catalog, New York Wire Railing Company (New York: Fowler & Wells, 1857). The company was later renamed the Composite Iron Works. See also the 1881 catalog of the J. L. Mott Company in New York for illustrations of cast-iron home and garden furniture.

6. "Art Reproductions in Cast Iron," *Furniture Gazette* (London) 3 (May 15, 1875): 623.

7. The first British iron frame was cast by Coalbrookdale in 1796 for the flax mill at Ditherington, Shrewsbury, still standing today. Cast iron was earlier used by Nash at Attingham in Shropshire in 1807.

8. Pugin, *True Principles*, 23–30. All subsequent quotations are from these pages.

9. Ruskin, *Complete Works* 8:81.

10. Ibid., 85–86.
11. Ibid., 16:385–386.
12. Ibid., 394.
13. Ibid., 8:83, 86.
14. Pugin, *True Principles*, 30.
15. "Exposition of British Industrial Art," 47.
16. "Illustrated Tour in the Manchester Manufacturing Districts: The Iron Works of Coalbrookdale," *Art-Union* 8 (Aug. 1846): 219. Subsequent references are to this page.
17. Ibid., 224. Subsequent references are to this page.
18. In actuality, Prussian foundries had been producing the identical items at least three years earlier. Coalbrookdale's cast-iron fruit dishes were also pirated copies of designs exhibited at Prussian industrial exhibitions in 1829 and 1831. My thanks to David de Haan, head of collections, Ironbridge Gorge Museum at Coalbrookdale, for information on Coalbrookdale's nineteenth-century castings.
19. Wyatt, "Iron-Work," 11–12.
20. Ibid., 74–75.
21. Ibid., 14.
22. John Haviland, *An Improved and Enlarged Edition of Biddle's Young Carpenter's Assistant* (Philadelphia, 1833–1837), 45, plate 60; *Miner's Journal*, quoted in *Hazard's Register of Pennsylvania* (July 1830), 400. For references to these works and a useful overview of cast iron's early history, see Turpin Bannister, "Bogardus Revisited," part 1, *Journal of the Society of Architectural Historians* 15 (Dec. 1956), 12–22. Bannister points out that during the 1840s, British and Belgian sheet-iron houses were noted in many American journals and that Bogardus "was almost certainly indebted" to British ironmaster William Fairbarn's prefabricated three-story iron building built in London, exhibited there, widely reported, and sent to Constantinople in 1840. The framework was entirely of cast iron (pp. 21, 15).
23. *New York Evening Post*, May 3, 1849.
24. See Bannister, "Bogardus Revisited," 19; the claim appears in James Bogardus, "Cast Iron Buildings: Their Construction and Advantages" (New York: J. W. Harrison, 1856), 3, reprinted in *The Origins of Cast Iron Architecture in America* (New York: DaCapo Press, 1970). On the verso side of the title page, Bogardus identifies "my friend Mr. John W. Thompson" as the author.
25. Bogardus, "Cast Iron Buildings," 4.
26. Ibid., 9. Interestingly, technological chauvinism continues into the twentieth century. In its garden catalog dating from the late 1980s, a New York cast-iron reproductions manufacturer, Irreplaceable Artifacts, advertised that its furniture, modeled from original nineteenth-century patterns, was cast in aluminum, "a material which is superior to cast iron because it is strong, easy to move, and will not rust."
27. Bogardus, "Cast Iron Buildings," 9.
28. Daniel D. Badger, *Illustrations of Architecture Made by the Architectural Iron Works of the City of New York* (New York: Baker and Goldwin, 1865), 4.

29. Ibid., 5.

30. "Architecture and Building Materials," *Scientific American* 1 (n.s., Nov. 26, 1859): 353.

31. Bishop, *History of American Manufactures*, 204, 205.

32. Ewing Matheson, *Works in Iron* (London: E. and F. N. Spon, 1873), 220, 221.

33. Ibid., 221–223.

34. Horace Greeley, Leon Case, et al., eds., *The Great Industries of the United States* (Hartford, Conn.: Burr, Hyde, 1872), 575–576.

35. Ibid., 379–383, 582.

36. Ibid., 382–384.

37. "Table Talk," *Appletons' Journal of Literature, Science and Art* 5 (Mar. 18, 1871): 323.

38. *The Engineer, Architect and Surveyor* (Chicago) 1 (May 15, 1874): 25.

39. *Appletons' Journal* 2 (Dec. 25, 1869): 601.

40. Ibid.

41. Matheson, *Works in Iron*, 283–284.

42. *Appletons' Journal* 5 (Mar. 18, 1871): 323–324.

43. John Claudius Loudon, *An Encyclopaedia of Cottage, Farm, and Villa Architecture and Furniture* (London: Longman, Rees, Orme, et al., 1833), 321.

44. *Art-Journal Illustrated Catalogue of the Great Exhibition*, 196.

45. *Art Journal* 5 (New York, 1879): 211.

46. For a close analysis of Victorian hallway furniture, see Kenneth L. Ames, "Victorian Hall Furnishings," *Journal of Interdisciplinary History* 9 (Summer 1978): 19–46.

47. Charles L. Eastlake, *Hints on Household Taste*, ed. with notes by Charles C. Perkins (Boston: James R. Osgood, 1876), 64.

48. Eastlake, *Hints on Household Taste*, rev. 2d ed. (London: Longmans, Green, 1869), 56.

49. Christopher Dresser, "Iron and General Hardware," in *Development of Ornamental Art in the International Exhibition* (London: Day and Son, 1862), 157.

50. Christopher Dresser, *The Art of Decorative Design* (London, 1862), 38.

51. Owen Jones, *The Grammar of Ornament* (1856; London: Day and Son 1868), 4–5.

52. Dresser, "Iron and General Hardware," 160.

53. Christopher Dresser, *Principles of Decorative Design*, 136.

6. CLASSICIZING THE MACHINE

1. Anna Seward, *Poetical Works*, ed. Walter Scott (Edinburgh, 1810), 2:315–316. The poem appears on 314–319.

2. See H. W. Dickinson, *A Short History of the Steam Engine* (Cambridge: Oxford Univ. Press, 1939), 73, 88.

3. For useful histories of the steam engine, see also H. W. Dickinson, *James Watt: Craftsman and Engineer* (Cambridge: Cambridge Univ. Press,

1936); George Watkins, *The Stationary Steam Engine* (Devon: David and Charles/Newton Abbot, 1968); T. E. Crowley, *The Beam Engine* (England: Senecio Press, 1962); and Carroll W. Pursell, *Early Stationary Steam Engines in America: A Study in the Migration of a Technology* (Washington, D.C.: Smithsonian Institution Press), 1969.

4. Thomas Tredgold, *The Steam Engine: Its Invention and Progressive Improvement* (1827); reprint, ed. W. S. B. Woolhouse, London: John Weale, 1838), 1:202.

5. Boulton and Watt Collection, portfolio 387, Public Libraries, City of Birmingham (England).

6. Thomas Tredgold, *Illustrations of Steam Machinery and Steam and Naval Architecture* (London: John Weale, 1840). This volume contained illustrations for Tredgold's *Steam Engine*, 1838 edition. The *Tiger* was designed by engineer Edward Bury of Liverpool.

7. Mumford, *Technics and Civilization*, 345, 332.

8. Herbert Read, *Art and Industry* (1953; reprint, Bloomington: Indiana Univ. Press, 1961), 121.

9. See Lionel Thomas Caswell Rolt, *A Short History of Machine Tools* (Cambridge, Mass.: MIT Press, 1965), 85, 89, 119, and Joseph Wickham Roe, *English and American Tool Builders* (New Haven, Conn.: Yale Univ. Press, 1916).

10. Rolt, *Short History*, 119.

11. Charles Dickens, *Hard Times* (1854; reprint, New York: Holt, Rinehart and Winston, 1964), 9.

12. Joseph Paxton, *The Engineer and Machinist and Engineering and Scientific Review* 2 (London, Dec. 1850): 312.

13. See Kouwenhoven, *Made in America*, for a discussion of the "vernacular tradition." Herwin Schaefer in *Nineteenth Century Modern: The Functional Tradition in Victorian Design* (New York: Praeger, 1970) has argued that there was a pervasive underground tradition of functionalism in Victorian design and cites examples of rational forms including Joseph Whitworth's screw-cutting machines, noted for their simplicity, precision, and direct approach to the design problem (p. 29).

14. Silliman and Goodrich, eds., *World of Science, Art, and Industry*, 12.

15. The Massachusetts manufacturer Chubbock and Campbell advertised a "Gothic Pattern" steam engine in 1844. In Britain, Marshall's Flax Mill in Leeds was ornamented with an Egyptian building facade and Egyptian motifs on the firm's double beam engine (1840). For illustrations of British-designed Gothic and Egyptian engines, see Pevsner, *High Victorian Design*, 24–25; see also Jerome Irving Smith, "Early Industrial Machinery and the Decorative Arts," *Art in America* 45 (Winter 1957–1958): 38–43.

16. Oliver Byrne, *The American Engineer, Draftsman and Machinist's Assistant* (Philadelphia: C. A. Brown & Co., 1853). The column, Byrne wrote, gave the engine a "compact and solid appearance" and provided a framework "free from all jarring or tremor, so common to most beam engines" (plate 6 caption, p. 66).

17. See *The Pneumatics of Hero of Alexandria*, ed. and trans. Bennet Woodcroft (London: Taylor, Walton and Maberly, 1851).

18. A useful study of American neoclassical architecture appears in Talbot Hamlin's *Greek Revival Architecture in America* (London and New York: Oxford Univ. Press, 1944). See also Andrew Jackson Downing, *The Architecture of Country Houses* (New York: D. Appleton, 1850). Examples of the "Grecian" style found in John Claudius Loudon's *An Encyclopaedia of Cottage, Farm and Villa Architecture and Furniture* (1833), a book widely read in England and America, were still included in the book's eleventh edition (1867).

19. Discussions of the Roman Revival and Thomas Jefferson appear in Marcus Whiffen, *American Architecture since 1780* (Cambridge, Mass.: MIT Press, 1969), 31–35. See also James S. Ackerman, *Palladio* (Harmondsworth: Penguin, 1967).

20. Thomas Webster, *An Encyclopaedia of Domestic Economy* (New York: Harper Bros., 1845), 247.

21. John Hall, *The Cabinet Maker's Assistant* (Baltimore: John Murphy, 1840). In England, an important source of late Regency patterns was Peter Nicholson and Michel Angelo Nicholson, *The Practical Cabinet Maker, Upholsterer, and Complete Decorator* (London: H. Fisher, 1826), which includes 81 plates of the classical order and ornament. For a historic tracing of Regency neoclassicism in patternbooks, see Steegman, *Victorian Taste*, 303–305.

22. Giedion, *Mechanization Takes Command*, 329.

23. John Ruskin, in *The Seven Lamps of Architecture*, scoffed at the historicized facades applied to the new railroad stations. What point was there in ornamenting stations, he wondered, when they were "the very temple of discomfort," where "people are deprived of that portion of temper and discretion which are necessary for the contemplation of beauty." Yet amidst his ridicule, Ruskin argued that the railroad station had a degree of legitimacy that precluded the need for ornamentation: "Railroad architecture has, or would have, a dignity of its own if it were only left to its work. You would not put rings on the fingers of a smith at his anvil." *Complete Works* 8:159.

24. Byrne, *American Engineer*, 66.

25. Samuel Clegg, [Jr.], *The Architecture of Machinery: An Essay on Propriety of Form and Proportion, with a View to Assist and Improve Design* (London: Architectural Library, 1842). Biographical information on Clegg is scarce. It is known that he worked as an engineer for several British railways, including the Great Western Railway, and was a professor of civil engineering at England's Putney College. His obituary appears in the *Minutes of the Proceedings of the Institution of Civil Engineering* 21 (1861–1862), 552–554.

26. Clegg, *Architecture*, 2, 1. Clegg's concerns about machine aesthetics also had an economic dimension, for as he wrote, poor design and a lack "of elegance in the contour of a machine is not only displeasing to the spectator, but disadvantageous to the manufacturer" (p. 2).

27. Ibid., 8.

28. Ibid., 2–4, 8.

29. Ibid., 3.

30. Ibid., 18.
31. Ibid., 33, 13.
32. Ibid., 14; Le Corbusier (Charles Edouard Jeanneret), *Vers une architecture* (Eng. version: Towards a New Architecture) (Paris: G. Crès, 1923), 106–107. For an extensive study of classicism and modernist design, see Rayner Banham, *Theory and Design in the First Machine Age* (London: Architectural Press, 1960; 2d ed., 1967).
33. Clegg, *Architecture*, 15–16.
34. Ibid., 13.
35. Ibid., 17, 8.
36. Wyatt, "Iron-Work," 10–11, 75.
37. Useful background on neoclassical-design sewing machines appears in Grace Rogers, *The Sewing Machine*, rev. ed. (Washington, D.C.: Smithsonian Press, 1976), 87–89; the decorative stenciling and lacquering of machines was described in the *Scientific American*, Aug. 20, 1881, cover story.
38. Kouwenhoven, *Made in America*, 22ff.
39. Kasson, *Civilizing the Machine*, 160.
40. *Manufacturer and Builder* 13 (Feb. 1881): 1.
41. Bishop, *History of American Manufactures*, 438–439.
42. "Universal Infidelity," 158–160.
43. Ibid., 159–160.
44. Byrne, *American Engineer*, 66.
45. "Form in Design," *Engineering* (London) 1 (Feb. 9, 1866): 85.
46. *Engineering* (London) 3 (Feb. 15, 1867): 150, 152; (Apr. 24, 1868): 397.
47. "Men and Things Mechanical," *Appleton's Mechanics' Magazine and Engineers' Journal* 1 (Nov. 1, 1851): 677.
48. Ibid., 678.
49. Ibid.
50. Horatio Greenough, *Form and Function*, ed. Harold A. Small (Berkeley and Los Angeles: Univ. of California Press, 1947), 22.
51. Horatio Greenough, "American Architecture," in *The Travels, Observations, and Experiences of a Yankee Stonecutter*, part 1 (New York: G. P. Putnam, 1852), 144. This essay was first published in *The United States Magazine and Democratic Review*.
52. Ibid., 138–139.
53. "Philadelphia Gas Works," in *Gleason's Pictorial Drawing Room Companion* 4 (1835): 216.
54. "Machine Tools at the Philadelphia Exhibition—No. 1," *Engineering* (London) 21 (May 26, 1876): 428.
55. Hiram S. Maxim, *My Life* (London: Methuen, 1915), 85–86.
56. "Machine Aesthetics," *American Machinist* 5 (Mar. 18, 1882): 8.
57. Ibid.
58. John H. Barr, "Aesthetics in Machine Design," *Cassier's Magazine*, Aug. 1892, 297.
59. Ibid., 298.
60. Ibid., 299.

61. Adolf Loos, "Glass and Clay" (1898), in *Spoken into the Void: Collected Essays 1897–1900*, intro by Aldo Rossi, trans. Jane O. Newman and John H. Smith (Cambridge, Mass.: MIT Press, 1982), 35.

62. Ibid., 36; Loos, "Review of the Arts and Crafts," in *Spoken into the Void*, 104.

AFTERWORD

1. "Industrial Design," *Engineering News* (Chicago), Feb. 19, 1876, 60.

2. Alfred Barr, Jr., foreword to *Machine Art*, exhibition catalog (New York: Museum of Modern Art, 1934).

3. In the computer-imaged landscape titled *Changing the Fractal Dimension*, produced by Richard Voss at IBM's Thomas J. Watson Research Center in Yorktown Heights, New York, 1983, a simulated mountainscape was set against a violet sky. By changing fractal dimensions, the blue-tinged mountains became craggy or were turned into hills. Voss's work is discussed and illustrated in Cynthia Goodman, *Digital Visions: Computers and Art* (New York: Abrams/Syracuse: Everson Museum of Art, 1987), 106, 114.

4. Oskar Schlemmer, diary entry, Sept. 1922, in *Letters and Diaries of Oskar Schlemmer*, ed. Tut Schlemmer, trans. Krishna Winston (Evanston, Ill.: Northwestern Univ. Press, 1990), 126.

5. See Nancy Burson, Richard Carling, and David Kramlich, *Composites: Computer Generated Photographs* (New York: Beech Tree/William Morrow, 1986) and *Faces: Nancy Burson*, exhibition catalog (Houston: Contemporary Arts Museum, 1992).

6. In *Turing's Man: Western Culture in the Computer Age* (Chapel Hill, N.C.: Univ. of North Carolina Press, 1984), J. David Bolter examines the impact of artificial intelligence research on human self-definition. He argues we no longer care about creating automatons that physically resemble human beings but are instead attempting, through the use of computers, to imitate human intelligence, and notes, "by making a machine think as a man, man recreates himself, defines himself as a machine" (p. 13).

7. Holzer's painting is illustrated in *Disarming Images: Art for Nuclear Disarmament* (New York: Adama Books, 1984).

8. A detailed description of Tinguely's sculpture is presented by Billy Klüver, an engineer for AT&T who assisted the artist in making the sculpture and who has long been associated with American art and technology projects. Klüver's reminiscence appears in K. G. Pontus-Hultén, *The Machine as Seen at the End of the Mechanical Age* (New York: The Museum of Modern Art, 1968).

9. Umberto Boccioni, Carlo D. Carrà, et al., "Technical Manifesto," Apr. 11, 1910, in *Futurist Manifestos*, ed. Umbro Apollonio, trans. Robert Brain (New York: Viking Press, 1973), 19–24.

10. A related image, American artist Gerald Murphy's painting *Watch* (1924–1925), is a yellow and gray abstracted view of watch cogs and wheels, but the painting suggests mechanism more than speed. For a discussion of Murphy's work, including his technological imagery, see William Rubin, *The Paintings of Gerald Murphy* (New York: The Museum of Modern Art, 1974).

11. American artist Joseph Stella visited Paris from 1911 to 1912, where he came in contact with Cubist and Futurist art, returning to produce his Futurist paintings of 1913–1916. Little-known American painter Frances Simpson Stevens (1881–1961) lived in Italy and exhibited at the International Futurists show held in Rome in 1914. British artist Christopher R. W. Nevinson was one of the only British artist converts to Futurism and a member of the Vorticists. For more on Stevens, see *Futurism and the International Avant-Garde*, exhibition catalog (Philadelphia: Philadelphia Museum of Art, 1980).

12. Cheney and Cheney, *Art and the Machine*, 97.

13. For studies of streamlining, see Jeffrey L. Meikle, *Twentieth-Century Limited: Industrial Design in America 1925–1939* (Philadelphia: Temple Univ. Press, 1979); Donald J. Bush, *The Streamlined Decade* (New York: George Braziller, 1975); and Richard Guy Wilson, "Transportation Machine Design," in *The Machine Age in America 1918–1941* (New York: Brooklyn Museum of Art/Harry Abrams, 1986), 125–147.

14. The painting was created for *Fortune*. The magazine's editors are quoted in Troyen and Hirshler, *Charles Sheeler*, 168.

15. Enrico Prampolini, "The Aesthetic of the Machine and Mechanical Introspection in Art" (1922; reprint, in *Machine-Age Exposition*, special edition of *Little Review* [New York], May 1927).

16. Barr, intro. to *Machine Art*.

17. Ibid., quoting Plato, *Philebus*, 51c.

18. Robert A. M. Stern with Raymond W. Gastil, *Modern Classicism* (New York: Rizzoli, 1988), 65, 76. Stern assumes that Moore's irony stems from his belief that the modernist architectural idiom—which was predicated on avant-garde, machine age technology—no longer makes sense in a world dominated by electronic communication technologies. For the modernist, although "modernism has been drained of most of its cultural meaning, it continues to perform syntactically." Moore's addition of classical design references can be viewed as an effort to add an element of meaning and continuity to a depleted modernist idiom (Stern, 65, 76). As another of the ironies associated with postmodernism, Moore's Piazza, with its classical imagery connoting permanence and stability, has fallen into a sad state of neglect and decay.

19. Charles Jencks in *Late Modern Architecture* (New York: Rizzoli, 1980) presents his 1978 essay mapping out the fundamental differences between late modern and postmodern architecture—late modern being an exaggeration of modernism, including "extreme logic" and making "mannered and decorative use of technology," postmodernism drawing on "historical memory, urban context, ornament, representation, metaphor," as well as "pluralism and eclecticism" (p. 13).

Index

Italic page numbers refer to illustrations or captions.

accidents: industrial, 193; railroad, 3–5, 15, 17, 32, 34, 48–57 *passim*, 64, 105, 192, 232 n33, 233 n49; steam boiler explosions, 16, *16*, 32–34, 40, 41, 44–46, 51; steamship, 15, 16, *16*, 34
Across the Continent: "Westward the Course of Empire Takes Its Way" (Currier and Ives print), 54
Adam, Robert and James, 118, 189
Adams, Charles Francis, 3, 16
AEG Turbine Factory (Behrens), 219
Aeolipile, 82
"Aerial Steam Carriage" patented by Henson, 90
aesthetics: Emersonian, 14; of handicraft, 26; of imitation, 20; Japanese, 144; of machine art, 204; of machine design, 19, 20, 26–29, 193–210, 217–219, 246 n26; of the picturesque, romantic, and sublime, 12; of the Podsnaps, 137, 138
"Aesthetics in Machine Design" (Barr), 207–208
Afternoon View of Coalbrookdale, An (Williams), *9*, 9–10
"Aims of Art, The" (Morris), 22
air brakes, 17, 52
aircraft design, 217
air pump, John Prince's, 184, *185*
Alken, Henry, 43

Allport, Sir James, 233 n48
aluminum, and technological chauvinism, 243 n26
American Engineer, Draftsman and Machinist's Assistant, The (Byrne), 186, *187*, 202
American Institute of Architects, cast iron debated at, 164
American Journal of Science (ed. Silliman), 130
American Machinist, The, 206
American Railroad Safety Appliance Act, 17
American Railway Scene at Hornellsville, An (Currier and Ives), 54, 232 n41
American Revolution, 228 n48
American System, 84, 236–237 n8. *See also* division of labor
Andrew Handyside and Company, 164
animation, computer, 213
anonymity, 122. *See also* depersonalization
Anshutz, Thomas, 35
anthropomorphism in machine design, 41, 82
Antiquities of Athens (Stuart), 189
anxiety: over counterfeit hollowware, 241 n40; about explosions, 32, 35, 39, 193, 200; rational response to, in Daumier, 61–63; about technology, 3–11 *passim*, 15–19, 31–34, 35, 70, 193, 200; travel, 18, 47. *See also* fear

Index

Appletons' Journal, 167–168, 169, 170
Appleton's Mechanics' Magazine and Engineers' Journal, 138, 203–204
aquatint, 12
Arc-et-Senans saltworks, 191
Architectural Iron Works, 163, 164
architecture: cast-iron facades, 21, 23, 113, 145, 153, 154, 161, 163, 164; Japanese, 28; modernist, 249nn18–19; neoclassical, 208–210; and ornament, 20, 23, 149, 151–153, *154*, 155, 193–198; postmodernist, 220–221, 249nn18–19; Shingle style, 28; and structural cast iron, 151, 159, 161–167, 177, 242n7
Architecture of Country Houses (Downing), 189
Architecture of Machinery (Clegg), 193–198
art: education, 229n55 (*see also* design: education); fine vs. decorative, 109–110; self-destroying, Tinguely's, 216; 20th-century, and the machine, 211–223
"Art and Craft of the Machine, The" (Wright), 28
Art and Industry (Greeley), 138–139
Art and Industry (Read), 183
Art and Photography (Scharf), 227–228n41
Art and the Machine (Cheney), 2, 217
art critics, 21–24, 25–29; on cast iron, 155–161; on electroplating, 134–144; on industrial design, 105–114
Art Education, Scholastic and Industrial (Smith), 116, 122
artificial intelligence, 248n6
"Artificial Man, The" (*Scientific American* essay), 87
Art Journal, 25, 107, 115, 122, 123, 126, 138, 141, 144, 171, 172, 176. *See also Art-Union*
art labor movement, American, *see* Smith, Walter
art metalwares, 129
art nouveau, 177
art objects, cast-iron, 149
Art of Decorative Design, The (Dresser), 175, 176

Arts and Crafts movement, British, 22, 113, 136, 209
"Arts and Crafts of Today, The" (Morris), 113–114
Art-Union, 105, 107–108, 110, 115, 119, 138, 145, 157–159
Ashbee, C. R., 113
atmospheric engine, 179
Atropos, 48
Atwater sewing machine, 199, *199*
authenticity and cast iron, 167–170
automatons, 18–19, 79, 81–84, 85, 89, 234–235n14; fear of becoming, 7
"Autoperipatetikos" doll, 84

Badger, Daniel, 145, 153, 163–164. *See also* Architectural Iron Works
"Bad Taste" (*Engineer, Architect and Surveyor* article), 168
Balla, Giacomo, 211, 212, 216
balloons, hot-air, 38, 44
Bannister, Judith, 236n9
Bannister, Turpin, 243n22
"Barge of Venus, The" (Middletown Silver Plate Co.), 122, *122*
Barr, Alfred Jr., 212, 218
Barr, John H., 207–208
Battle of Lights, Coney Island (Stella), 1
Baudrillard, Jean, 19
Bauhaus, 121, 214
beam engines, *180*, 186, *188*, 202, 205
Beattie, William, *127*, 128
bedsteads, cast-iron, 149
Behrens, Peter, 219
Bell, John, 120, 130, 150, *151*
Bell, Quentin, 120
"Bell-Tower, The" (Melville), 85
Benjamin, Walter, 227n41
Bergson, Henri, 88–89
Berlin, Prussian royal iron foundry at, 147–149, 242n4
Bessemer converter, 73, 74, *75*, 76
Best Friend of Charleston, locomotive, 34, 46, 51, 232n33
betterment of self, 228n48
Bibliothèque Sainte-Geneviève, 152
bicycle, 209, 210
Biddle's Young Carpenter's Assistant, Haviland's edition of, 161
Biggs on Artificial Limbs, 87

Birmingham, 108, 109, 131, 132, 133, 136, 139, 178
Bishop, John Leander, 139, 164, 200–201
Bizzarie di varie figure (Bracelli), 89
blacks, segregated railway seating for, 60, 233n44
Blacksmith's Shop, The (Wright), 78
Blair, Francis, 226n30
Blake, William, 12, 31
Blast Furnace at the West Point Foundry (Chapman), 13
"Body-Fanner, Nut-Cracker, and Wine Helper" (Seymour), 92, *93*
Bogardus, James, 21, 145, 152–153, 161, 162–163
boilers: anthropomorphic, 82; exploding, 33–34, 40, 41, 44–46, 51; laws regulating, 33; sectional, 231n21
Bolter, J. David, 248n6
Boston, art education in, 121
Boston and Lowell Railroad car house, 191–192
Botanic Garden, The (Darwin), 7, 37
botany, 111, 116, 118, 123, 144, 175, 176
boudoir, machines tamed for, 200
Boulsover, Thomas, 131
Boulton, Matthew, 37, 40, 105, 109, 131, 179, 182, 236n8, 236n9
Boulton and Watt, 151, 179, *180*, 182, 205
Bourne, Thomas Cooke, 54
Bracelli, Giovanni Battista, 89
braking systems, railway, 17, 52
Branca, Giovanni, 82
breaking frame: industrial, 6, *6*, 11, 42; perceptual, 3–4, 6, 15–16, 17, 46, 61
Breck, Samuel, 47
Brighton Royal Pavilion, 152
britannia metal, 131, 132, 240n10, 241n15
British Reform Bill, 230n15
bronze, imitation of, 147, 149
Brookes's Flax Mill, explosion at, 33
Brooklyn Navy Yard, drydock pump at, 205
Brown, Richard, 228n48
Brunel, Isambard Kingdom, 55
Brunel, Marc Isambard, 55, 184
Brunelleschi, Filippo, 130

Buena Vista, locomotive, 34
building facades, cast-iron, 21, 23, 113, 145, 153, *154*, 161, 163, 164
buildings, prefabricated, 242n22
Bureau of Education report on designer training, 117, 121
Burke, Edmund, 12, 225n22
Burlington, Richard Boyle, 3d earl of, 190
Burson, Nancy, 215
Bury, Edward, *183*, 245n6
Bury, Thomas, 54–55, *192*
Byrne, Oliver, 186, 202, 245n16

Cabinet Maker's Assistant (Hall), 190
calling-card basket, iron, 159
calling-cards, cast-iron, 149
calotype, 21
Caricature, La (weekly), 56, 57
caricaturists, *see* satiric artists
Carlyle, Thomas, 18, 31, 79–81
Carron Company, 119, 147, *148*
Carson Pirie Scott store, 177
"Carving by Machinery" (*Art-Union*), 110
Cassier's Magazine, 207
cast iron, 106, 109, 112, 115, 135, 137; American critical response to, 161–167; architectural, 21, 23, 113, 145, 161–167, 177, 242n7; art reproductions in, 149, 151, 166; British critics of, 155–161; development of, 145–149; fruit dishes, 243n18; furniture, 149, *150*, 170–176; housewares, 21; Iron Bridge near Coalbrookdale, 224n16; as primary art medium, 149–151; prefabricated buildings, 242n22; Prussian, 145, 147–149, 242n4, 243n18, 243n22; upstaged by aluminum, 243n26
Cast Iron Buildings: Their Construction and Advantages (Bogardus), 162
catalogs: *Art Journal*'s, of London Crystal Palace Exhibition, 138, 171; Coalbrookdale Company's, 145, 168, 172, 176; Machine Art show, Museum of Modern Art, *219*; of New York Exhibition of 1853, Silliman's, 186; New York Wire Railing Company, 153
Catch me who can, locomotive, 30
celluloid, 214

254 Index

Centennial Exhibition of 1876, Philadelphia, 15, 73, *74*, 124, 128, 140, 141, 206
Cham (Amédée de Noé), 98–99, *99*
Changing the Fractal Dimension (Voss/IBM), 248 n3
Chaplin, Charlie, 227 n39
Chapman, John Gabsby, 13
Charivari, Le, 226 n29
Charleston Mercury, 34
chauvinism, technological, 146, 162, 166–167, 221
Chemins de Fer series (Daumier), *32*, 57, *58*, 58–59, *59*
Cheney, Sheldon, 2, 217
chiaroscuro, 78
Chicago Stock Exchange building, 177
Chicago waterworks stations, 208
"Choix du Wagon, Le" (Daumier), *32*
chromolithography, 54, 223 n7
Chubbock and Campbell steam engine, 245 n15
Civilization and Its Discontents (Freud), 90
Civilizing the Machine (Kasson), 24
Clarke, Isaac Edwards, 121
class, social, 137, 157, 160–174 *passim*, 194, 200, 228 n48; aesthetic criticism and, 22–23, 137; and railway travel, 59–60, 62–66, 232–233 n44, 233 n46, 233 n48
Classicism, "Ironic," 220, 249 n19. *See also* neoclassicism
Classic Landscape (Sheeler), 1, *2*, 212
Clegg, Samuel [Jr.], 193–198, 207, 210, 217, 246 nn25–26
clock, as metaphor, 224 n13
Clock, The (Goncharova), 217
clockmaking, 82, 183
coach seating, railroad, 233 n46
Coalbrookdale, 8–10, 224 n16; School of Art at, 120–121; in Seward's "Coalbrook Dale, " 178
Coalbrookdale by Night (de Loutherbourg), *11*, 11–12, 211, 225 n22
Coalbrookdale Company, 115, 119, 123, 146, 150, 157, 158, 178, 242 n7; art objects by, 150–151, *151*; catalogs, 145, 147, 168, 172, 176; designers for, 151–152, 174–176 (*see also* Dresser, Christopher)
"Cog, The" (Kupka), 101

coke, 225–226 n25
Cole, Henry, 22, 25, 116, 117, 120, 121, 136, 201, 238 n29
Colfax, Schuyler, 226 n30
Colman, Samuel, 14, *15*
Columbian Exposition, Chicago, 208
columns, neoclassical, 153, 161, 162, 163, *165*, 180–187 *passim*, 195–198, 202, 205–206, 220, 221, 245 n16
comfort, 64–66, 223 n7
Comforts of a Cabriolet (Cruikshank/Edgerton), 44
Composite Iron Works, 242 n5. *See also* New York Wire Railing Company
computer-assisted imagery, 212–214, 215, 248 n3
Condillac, Etienne Bonnot de Mably de, 86
"Conducteur!" (Daumier), *59*
Connecticut River Steamboat Company, 33
Consolidated Gas Company meter, *76*, 76–77
Consort of Taste (Seegman), 228 n50
copies: aluminum, of 19th-century cast iron, 243 n26; of artworks, in cast iron, 149, 151, 166; through electrotyping, 112, 130, 134, 138. *See also* imitative arts
Corliss engine, 73–74, 206
Cottingham, Lewis, 190
craftsmanship, 113, 114. *See also* Arts and Crafts movement; handicrafts
critics, 21–24, 25–29; on British cast iron, 155–161; of electroplating, 134–144; on industrial design, 105–114
Crooke, Charles, 157
Cropsey, Jasper F., 227 n31
crowding, 43
Cruikshank, George, 42–43, 44, 88
Cruikshank, Robert, 42
Crystal Palace: London's, 69, 72, 152, *153*, 185; New York's, 129, 166, 170
Crystal Palace exhibitions, *see* Great Exhibition, London; New York Exhibition of 1853
Cubism, 249 n11
Cugnot, Nicolas, 39–40
Currier and Ives, 13, 14, 55–56, 226 n30

Dada, 2, 94, 215
Daguerre, Louis, 21, 227n41
Dance, Sir Charles, 41–42
"Danger Ahead . . ." (*Harper's* illustration), 54
"Danger Signal on the Erie Railway . . ." (*Leslie's* illustration), 17, *18*
Darby, Abraham, 8, 146, 224n16
"Darktown" series (Currier and Ives), 55
Darwin, Erasmus, 7, 10, 37, 178
Daumier, Honoré, 31, *31*, 56, 57–64, 88, 98, 99–100, 226n29, 233n49
David, Theodore R., 73, *74*
"David and Sampson," ironworks engine, 186
da Vinci, Leonardo, 100–101, 110
DC-1, -2, and -3, Douglas aircraft, 217
"Death and His Brother Sleep" (Tenniel), 233n49
decals, 199
deceit, 129, 142, 162, 194; in cast iron, 155; in hollowware, 241n40; "operative," 111, 156
Dederick, Zadoc P., 83
De diversis artibus (Theophilus), 131
de Dunin, Count, 69
deerhound table, Bell's cast-iron, 150, *151*
"Delicatessen Order," architectural, 220
de Loutherbourg, Philip James, 11–12, 211, 225n22
Delteil, Loys, 233n49
democratization of elite norms, 228n48
Department of Practical Art, School of Design (Britain), 120, 121
Department of Science and Art, School of Design (Britain), 120
Department of the Interior, U.S., *see* Bureau of Education report
depersonalization, 61, 98, 114, 215. *See also* mechanized humans
Descartes, René, 81
design: critical skepticism, 105–114 *passim*, 133–144 *passim*; education, 116, 117, 120–121, 229n55; piracy, 243n18; vernacular tradition, 245n13 (*see also* handicrafts). *See also* Dresser, Christopher; Jones, Owen; Smith, Walter
Deutscher Werkbund, 121

DeWitt Clinton, locomotive, 46
Dibdin, Charles, 12
Dickens, Charles, 22, 48–49, 121, 137, 184, 227n39
die-stamping, 130–131, 132, 135, 236n9
dining cars, railroad, 64, 233n46
Discourse on Method (Descartes), 81
dishonesty, *see* deceit
disorientation, 4, 43, 46, 47. *See also* anxiety
division of labor, 22, 84, 109, 236n8. *See also* American System
Dixon, James, 123
Dixon and Sons, Sheffield, 137, 144
Dr. Strangelove (Kubrick), 50
dollmaking, 235n18
dolls, mechanical, 84; talking, 84–85
Dombey and Son (Dickens), 48
dome, cast-iron, Coalbrookdale's, at London Crystal Palace Exhibition, 150
double, automaton as, 19
Douglas aircraft of 1930s, 217
Downing, Andrew Jackson, 189
Dresser, Christopher, 26, 27, 112, 117, 119, 123–125, *143*, 143–145, 168, 175–176, 242n43
drop press, 112–113
Duchamp, Marcel, 215
duck, automated, 82
Dyce, William, 117
Dyer, John, 7–8, 224n10

Eagle Slayer, The (Bell), 150
Eastlake, Charles, 25, 137–138, 172, 174
Edgerton, M., 44
Edinburgh Review, 79
Edison, Thomas, 84
"Effects of the Turnstile on Crinoline Petticoats" (Daumier), 99–100
Egyptian revival, 190, 245n16
Eidophusikon, de Loutherbourg's, 12–13, 225n24
electrobronzing, 147
electrometallurgy, 239n5. *See also* electroplate; electrotyping; silverplate
electroplate, 21, 109, 126–134 *passim*, 176, 240n10; critical response to, 134–144; flatware, 240–241nn14–15; hollowware, 240–241n15, 241n40;

electroplate (*continued*)
  touchmarks, 240n15, 241n36, 241n40
electrotyping, 106, 109, 112, 129, 130, 138, 239–240nn5–6
elegance, inner, 203
Elementary Principles of Carpentry (Tredgold), 181
Elements of Electrometallurgy (Smee), 130
elevation/degradation, 108, 138–144 *passim*, 229n52. *See also* class, social; status
Elijah of Chelm, 85
Eliot, George, 105
Elkington, George, 21, 126, 130, 131, 132
Elkington, Henry, 131, 132
Elkington, Mason & Co., 109, 119, *127*, 130, 132, 133, 140
Ellis, Edward, 83–84
Elopement Extraordinary (satiric print), 50, *51*
Emerson, Ralph Waldo, 14
Encyclopaedia of Cottage, Farm, and Villa Architecture and Furniture (Loudon), 171, 246n18
Encyclopedia of Domestic Economy (Webster), 190
Engineer and Machinist, The, 185
Engineer, Architect and Surveyor, The, 167–168
Engineering, "Form in Design" appears in, 202
Engineering News, 27, 212
Engineering Reminiscences (Porter), 74
engravings: electrotype duplication of, 138; satiric, 230n15
"Entrance to the Large Tunnel, The" (Daumier), 60
Erie Railway, 17, 54
Essay on Man (Pope), 8
Essay on the Picturesque, An (Price), 225n22
etchings, satiric, 230n15
Etruria ceramic works, *see* North Staffordshire works
"Etruscan" style, 134
Euston Station, *192*
Evening Post (New York), 161
Excursion, The (Wordsworth), 10

exhibitions: of American art about machines, 218; Chicago Columbian Exposition, 208; London "Crystal Palace" (*see* Great Exhibition, London); New York "Crystal Palace" (*see* New York Exhibition of 1853); International Exposition (London, 1862), 175; Paris, of 1855, 98–99, 152; Paris, of 1878, 117; Prussian, 243n18
"Expanding Figure of a Man" (de Dunin), 69–71, 80, 94
Experiment, locomotive, 48
Experiment on a Bird in the Air Pump (Wright), 107
explosions, 92, 179, 211–212, 216, 231n21, 232n33; at Coalbrookdale foundry, 10, 224n17; boiler, 16, *16*, 32–34, 40, 41, 44–46, 51; fear of, 32, 35, 39, 193, 200
Exposition Universelle (Paris, 1855), 98–99, 152

facades, building: cast-iron, 21, 23, 113, 145, 153, *154*, 161, 163, 164; neoclassical, for railway stations, 246n23
fashion, in female dress, 99–100
fear: of assault while traveling, 62; of explosions, 32, 35, 39, 193, 200; of violence from machine, 96, 98. *See also* anxiety
Felix Holt, the Radical (Eliot), 105
female identity, 215
"Few Small Inconveniences, A" (Seymour), 90–91, *91*
Figaro (satirical magazine), 45
finishes: for cast iron, 147, 167–170; japanned, 199. *See also* electroplate; veneer
first class, rail travel by, 64–66. *See also* luxury railroad cars
First Class Carriage, The (Daumier), 63–64
First Class: The Meeting . . . And at First Meeting Loved (Solomon), 233n50
Fisher, Marvin, 229n52
flatware, electroplated, 240–241nn14–15
Flaxman, John, 118
Fleece, The (Dyer), 7–8
Flight of Intellect, The (satiric print), 49, *50*

Florence (Italy) Baptistery doors, 130
flutist, automated, 82
flyball, *see* governor
fly press, 109
Ford Motor Company River Rouge factory, 1
Forging the Shaft (Weir), 77
"Form and Design" (*Engineering* article), 202
Fortune magazine, 217
Foundry (Cropsey), 227n31
fractal dimensions, 248n3
fragmentation/fracturing, 3, 5, 20, 29, 108, 211; between mechanistic and organic, 111; of designer-manufacturer relationship, 106–107, 118
frame: artistic, first-class railroad compartment as, 65–66; perceptual, 3–4, 6. *See also* breaking frame
Frame Analysis (Goffman), 3
Frankenstein (Shelley), 85–87; monster in, 69, 89, 215, 218
Frank Leslie's Historical Register, 139–140
Frank Leslie's Illustrated Newspaper, 17, *18*
Frank Leslie's Weekly, 234n51
Franklin Institute committee on steam boiler explosions, 33
Franklin Ironworks (Philadelphia) beam engine, 186, *187*
Frederick the Great, 147
French Impressionists, 66
Fulham Gasworks gasholder, 202
Fulton, Robert, 30, 33
functionalism, 182, 203–204, 207, 209, 245n13
furniture: cast-iron, 149, *150*, 170–176; neoclassical, 190, 246n21
Futurism, 1, 211, 216–217, 249n11

Galvani, Luigi, 85
Gare de l'Est, 152
Garrick, David, 12
gasholder, Fulham Gasworks, 202
Gaynor, J. P., 153, *154*, 167
generator, steam-powered, for electroplating, 132
German silver, *see* nickel silver
Germany: Bauhaus, 121, 214; design quality, 123; Deutscher Werkbund, 121; Prussian cast iron, 145, 147–149, 242n4, 243n18, 243n22; Weimar Republic artists, 94, 214
Giedion, Siegfried, 191, 238n29
Gifford, Sanford R., 14
giganticism of machines, 7, 15, 17, 68–79, 178. *See also* scale
gig mill, 11
Gillray, James, 36, 42
Gilpin, Rev. William, 225n22
"Glass and Clay" (Loos), 209
glass house, 186–187
Gleason's magazine, 206
Gleiwitz (Gliwice), Prussian royal iron foundry at, 147–149, 242n4
God (Schaumberg), 212
Goffman, Erving, 3
Going It by Steam (Seymour), *45*, 45–46, 92
Goldberg, Rube, 92
golem legends, 85
Goncharova, Natalia, 217
Gordon, Sir Willoughby, 41
Gorham Company, 133, 140, 142, 221
Gothic revival, 110, 135, 171, 190, 245n15
governor, steam engine's, 179, 180
Graham, Charles, 73, *75*, 76
Grammar of Ornament (Jones), 118, 144, 176, 242n43
Grant, Ulysses S., 226n30
Graves, Jean, 101
Graves, Michael, 28
Great Exhibition, London (1851), 69–71, 72–73, 110, 116, 126, 130, 149, 150, 151, 171, 201, 238n30; *Art Journal's* catalog of, 138, 171
"Great Gas Meter of the Consolidated Gas Company of New York, The" (*Scientific American* illustration), 76
Great Industries of the United States (Greeley), 166–167
Great Western Railway, 34, 216, 246n26
Grecian revival, 246n18. *See also* neoclassicism
Greeley, Horace, 112, 129, 138–139, 140, 164, 166–167
Greenough, Horatio, 204–205, 217–218
Grosz, George, 214
Gun Foundry, The (Weir), 77, 77–78, 79
Gurney, Goldsworthy, 38, 40, *40*

Haber, Henri, 27
Hagley Temple (Stuart), 189
Hague Street explosion (New York City), 33
Hall, John, 190
hallmarks, misleading, 132
hallway furniture, cast-iron, 172, *173, 174, 177*
handicrafts, 26, 111, 135, 236n8, 237n20. *See also* Arts and Crafts movement
Handyside and Company, 164
Hard Times (Dickens), 121, 184, 227n39
Hargreaves and Company, 202
harmony, 207–208
Harper Brothers Building, New York, 153
Harper's New Monthly Magazine, 101, *102*, 133, 141–143
Harper's Weekly, 13, 15, 16, 32, 53–54, 55, 64, 73–74, *74*, 75, 76
harpsichord-playing automaton (Jacquet-Droz), 82, *83*
Harrison Street Waterworks Station, Chicago, 208
Haughwout Building, New York, 153, *154*, 167
Haviland, John, 161
Heath, William, 36, 37–39, *39*, 42, 43, 88
hell, industrial images of, 9, 12, 31, 43, 60, 106, 226n31
Henson, William, 90
Hephaestus, 82
Herald Tribune (New York), 112, 129, 140
Herculaneum excavations, 187
Hero of Alexandria, 82, 187
Heskett, John, 236n8
"Hindrances to the Progress of Applied Art" (Dresser), 26
Hine, Lewis, 102–104
Hints on Household Taste (Eastlake), 25, 137–138, 172, 174
History of American Manufacturers (Bishop), 139, 164, 200–201
Hoffmann, E.T.A., 85
Hogarth, William, 36
holloware, electroplated, 240–241n15, 241n40
Holzer, Jenny, 216

Homage to New York (Tinguely), 216
home electroplating of plants and animals, 133
Homeric automatons, 82
Homme machine, L' (La Mettrie), 81, 234–235n14
honesty in design, 136, 168. *See also* deceit
Hope, Thomas, 190
horse-drawn carriage, 42–43
Horses 'Going to the Dogs' (Cruikshank), 43
Household Furniture and Interior Decoration (Hope), 190
House of Commons report on steam safety, 231n21
Howells, William Dean, 73
Huckleberry Finn (Twain), 35
Hudson River School, 230n14
Hughes, Hugh, 4, *5*, 36
Hukin and Heath, *143*, 144
human contours, vs. machines, 94–100
Humorous Sketches (Seymour), 45
"Humors of Railroad Travel" (*Harper's* series), 55
Huskisson, William, 4, 49
Hyde Park As It Will Be (Leech), 43, *44*

"idealists" vs. "utilitarians," 198
identity: Arcadian, 158; and artificial intelligence, 248n6; female, 215. *See also* depersonalization
Iliad, The (Homer), 82
Illustrated Cyclopedia of London Crystal Palace Exhibition, 70
Illustrated Exhibitor, 132
Illustrated London News, 15
imitations: building facades, cast-iron; 21, 23, 113, 145, 153, *154*, 161, 163, 164; copies of artworks, 112, 130, 134, 138, 149, 151, 166; mass-produced, 20; of bronze, 147, 149; of hand work, 156 (*see also* handicrafts); of marble, 160; of solid silver (*see* silverplate)
imitative arts, 106–125 *passim*, 221. *See also* cast iron; electroplate
Impressionists, French, 66
"Impressions et Compressions de Voyage" (Daumier), *58*
Industrial Exhibitions: International,

London (1862), 175; Paris, of 1855, 98–99, 152; Paris, of 1878, 117; Prussian, 243n18. See also Centennial Exhibition of 1876; Great Exhibition, London; New York Exhibition of 1853
"Industrial Palace in Hyde Park, The" (Paxton), 185–186
inferno, industrial images of, 9, 12, 31, 43, 60, 106, 226n31
Inness, George, 35, *36*, 230n14
Inside of a Smelting-House at Broseley, The (Robertson), *68*, 68–69
Ins Leere gesprochen (Loos), 209
"Instructions for the Multiplication of Works of Art" (Spencer), 130
integration of technology, 20, 24–25, 29, 41, 108. See also legitimization of technology
Interior, U.S. Dept. of, *see* Bureau of Education report
International Business Machines' and Voss's fractals, 248n3
International Exposition (London, 1862), 175
International Style (architecture), 219
intimacies, illicit, on railroad, 65–66, 234n51
Iron Bridge, Severn River, 224n16
Iron Forge Viewed from Without, An (Wright), *78*, 78–79, 107
"Ironic Classicism," 220, 249n18
"Iron-Work and the Principles of Its Treatment" (Wyatt), 115, 159–161
ironworks technology, 225–226n25. See also individual ironworks
Italian Futurism, 1, 211, 216

Jacobi, Moritz Hermann von, 129, 239–240n5
Jacques, William W., 84
Jacquet-Droz, Pierre and Henri-Louis, 82, 85
James Dixon and Sons, Sheffield, 137, 144
James L. Pettigru, locomotive, 34
James Watt and Company pumping engine, 205
Japan: Its Architecture, Art and Art Manufacture (Dresser), 144
Japanese architecture, 28

Jeanneret, Charles Edouard (Le Corbusier), 197, 204
Jeannest, Pierre-Émile, 119
Jefferson, Thomas, 190
Jencks, Charles, 249n19
Jennings, Humphrey, 225n24
"Jerusalem" (Blake), 12
Johnson, Philip, 219
Jones, Owen, 118, 136, 144, 176, 238n29
Jordan, C. J., 110, 130, 239n5
Journal of Design and Manufactures, 22, 25, 27, 116, 120, 137, 159–161, 198, 201
Journal of the Society of Arts: Dresser article in, 123–124
J. T. Sutton Company's beam engine, 186–187

Kasson, John, 24
Kemble, Fanny, 31, 48
Kennedy cotton-spinning factories (Manchester), 182
Kirkham, Thomas N., 202
Kiss in the Dark, A (Currier and Ives), 55–56
kite-propelled vehicles, 38
Klüver, Billy, 248n8
Kouwenhoven, John, 24, 27, 186, 200
Kubrick, Stanley, 50
Kupka, Franz, 101

Labrouste, Henri P. F., 152
Lachhammer foundry (Germany), 242n4
Lackawanna Valley, The (Inness), 35, *36*
Lady Pink, 216
Lafever, Minard, 189
Laing stores (New York), 161
La Mettrie, Julien Offray de, 234–235n14
"Lamp of Truth, The" (Ruskin), 136, 156
Lancashire, antimachine riots in, 225n21
Landscape with Stottsville in the Distance (Gifford), 14
landscapes: computer-assisted images of, 212–214, 248n3; painting of, 9–15 *passim*, 34–36, 86
Lardner, Dionysius, 52

"Lathes, The" (*Harper's* illustration), 101, *102*
Laughter (Bergson), 88–89
Laws of Fiesole, Ruskin preface to, 121
Le Corbusier (Charles Edouard Jeanneret), 197, 204
Ledoux, Claude-Nicolas, 191
Leech, John, 43, *44*
Leeds, antimachine riots in, 225n21
Léger, Fernand, 214
legitimization of technology, 7–8, 27, 34, 114–125, 159–172 *passim*, 189–193, 246n23
Leibniz, Gottfried Wilhelm, 224n13
lithographs, 13, 37, 56, 230n15. *See also* chromolithography
Liverpool and Manchester Railway, 3–4, 31, 46, 49
Living Made Easy series (Seymour), 92, *93*
Locomotion (Seymour), 19, 79, *80*, 83, 90, 92; second version, 90–91, *91*, 92
locomotives, experimental, 30, 180. *See also individual engines*
Loew, Rabbi of Prague, 85
Loewy, Raymond, 217
London Observer, 34, 38, 40–41
London *Times*, 30, 201–202
Looking Glass, The (monthly), 37
looms, power, 11
Loos, Adolf, 209–210
Loudon, John Claudius, 171, 246n18
Lowell Institute, Boston, 121
Luddites, 11, 42
luxury railroad cars, 64, 233n46

McAlpine, William, 205
machine: aesthetics of, 19, 26–29, 178–207, 217–219, 246n26; anthropomorphic, 41, 82; destructiveness of, feared, 61, 229n52 (*see also* explosions); encircling, 100–104; "in the garden," 14, 35; and human contours, 94–100; legitimization of, through design, 114–125, 144; as model for humans, 248n6 (*see also* mechanized humans); as monster, 84–88; runaway, 17, 30, 32, 34, 39, 46, 57, 58, 60, 179, 193; in 20th-century art, 211–223; violent death by, 61 (*see also* explosions). *See also individual devices*
"Machine Aesthetics" (*American Machinist* article), 206–207
Machine-Age Exposition (New York, 1927), 218
machine art: aesthetic, 204; Museum of Modern Art show (New York, 1934), 218, *219*
machine gun, Maxim's, 206
machine tools, 199–200
Machine Tournez Vîte (Picabia), 215
McKim, Mead and White, 191
McLean, Thomas, 37, 44
"Madame Deménage" (Daumier), 233n49
Made in America (Kouwenhoven), 24
"Making Bessemer Steel at Pittsburgh" (Graham), 73, *75*, 76
Manchester Exhibition of Industrial Arts, 145, 157
Manchester textile trade, 139
Mantoux, Paul, 224n10
March of Intellect, The (Seymour), 92, 94, *95*
March of Intellect, The, series (Heath), 38–39, *39*, 43, 92–94
Marryat, Frederick, 47
Marshall's Flax Mill, 245n15
Marx, Karl, 18
Marx, Leo, 14, 35, 226n31
Massachusetts Institute of Technology, 121
Massachusetts Teachers' Association, 117
mass production, 236n8. *See also* American System
mastery: of technology, in satiric prints, 49–50, *50*, *51*; threatened by technology, 47
Matheson, Ewing, 164–166
Maudslay, Henry, 184
Maxim, Hiram, 206
Mechanical Abstraction (Schamberg), 212
Mechanics' Magazine, 3–4, 31
mechanization, attractive to U.S. industry, 226n27
mechanized humans, 49, 84–87, 214, 215; fear of becoming, 7, 79–81, 87–88. *See also* automatons

Melksham, steam carriage mobbed at, 42
Melville, Herman, 85, 100
Memorial of Horatio Greenough, A (Greenough), 204
"Men and Things Mechanical" (*Appleton's* article), 203–204
Mène, Pierre-Jules, 150
Meriden Britannia Company, 133, 134, *135*, 240n14
metonymy in neoclassical design, 191, 198, 220
Middletown Silver Plate Company, 122, *122*
Midland Railroad Company, 233n48
Mies van der Rohe, Ludwig, 219
Milhau drugstore (New York), 161
Miller, Henry (composer), 91
Mills, Robert, 189
Milton, John, 12, 225n24
Miner's Bank (Pottsville, Pa.), 161
modernism, 144, 204, 217–219, 221, 249n18
"Modern Manufacture and Design" (Ruskin), 107, 228n42
Modern Painters (Ruskin), 111
Modern Times (Chaplin), 227n39
Mohawk and Hudson railroad line, 46
Montgolfier, Joseph-Michel and Jacques-Etienne, 38
Monticello (Jefferson), 190
monumental, the, 225n22. *See also* giganticism of machines
Moore, Charles, 28, *220*, 220–221, 249n18
Morel-Ladeuil, Léonard, 119
Morning View of Coalbrookdale (Williams), 9
Morris, William, 21, 22, 23, 109, 113–120 *passim*, 136, 176, 209
Morrison, Enoch Rice, 84
Morrison, James, 120
Mott, J. L., 149
Moyr-Smith, J., 119, 175
Mumford, Lewis, 71, 182–183, 184
Murdock, William, 179–180
Murphy, Gerald, 248n10
Museum of Modern Art (New York), 212, 216, 218
museum reproductions, *see* replicas
musicians, mechanical, 82, *83*

Nash, John, 152, 189
Nash, Joseph, 153
Nasmyth, James, 71–72
"naturalism" in design, Victorian, 134, 135
"Nature of Gothic, The" (Ruskin), 100, 114
neoclassicism, 134, 153, 246n18, 246n21; classical antecedents, 187–190; in cast iron, 160–161; in Expanding Man, 71; in machine design, 24–25; in steam-engine design, 180–207 *passim*; in stove design, *148*, 171. *See also* "Ironic Classicism"
Nevinson, Christopher, 217, 249n11
Newcomen, Thomas, 179
New England steamboat explosion, 33
New Principles, or the March of Invention (pub. McLean), 44–45
Newtonian universe, 8, 81, 224n13
New Travel on a Steam Engine (Barth), 49
New York Central Railroad, 217
New York Exhibition of 1853, 15, 112, 129, 130, 139, 140, 149, 152, 186
New York Wire Railing Company, 149, *150*, 153, 172, 242n5
Nicholson, Peter and Michel Angelo, 246n21
nickel silver, 132, 240n10
Noé, Amédée de, *see* Cham
norms, democratization of, 228n48
North-Side Station, Chicago waterworks, 208
North Staffordshire works, Wedgwood's, 105, 108
nuclear war, 215–216

Off His Nuts (Currier and Ives), 55
Oldsmobile "Rocket," 50
Opening the Great Exhibition: The Foreign Nave (Nash), *153*
operative deceit, 111, 156
organic forms, 134. *See also* botany
ornament, 20; architectural, 20, 23, 24, 149, 151–153, *154*, 155, 193–198, 246n23; *Art-Union* critiques cast-iron, 158–159; delicacy and precision in, 165; excessive, 137, 155; on machinery, 24–25, 178–210 *passim*, 229n52; vs. utility, 201–202

"Ornamentation Considered as High Art" (Dresser), 123
ornithopters, 43, 101
Our Mutual Friend (Dickens), 22, 137

Paddington Station, 152, 233n49
Paestum, excavations at, 187–188
painting: cast iron finished with, 167–170 *passim*; of industry in landscape, 1–2, 9–15 *passim*, 34–36, 86. See also Cubism; Dada; modernism; postmodernism; Precisionism
paktong, *see* nickel silver
"palace cars," 64
Palladian motif, 190, 195
Palladio, Andrea, 189–190
Pandemonium, 12–13, 225n24; "talking" doll factory as, 85
Paracelsus, 85
Paradise Lost (Milton), 12, 225n24
Paris Expositions: of 1855, 98–99, 152; of 1878, 117
Paris–St.-Germain railroad line, 46
Paris–Versailles railroad line, 1842 accident on, 56–57
Parliament Select Committee on art education, 120
parlor cars, railroad, 60, 233n46
Parthenon, 195, *196*, 197
pattern books, 118, 147, 190, 246n21
pattern makers, 147
Paul, Lewis, 8, 224n10
Paxton, Joseph, 69, 72, 152, 185–186
Pennsylvania Railroad Station, New York, 192
Perkins, Charles, 172
petticoats, 99
Pevsner, Antoine, 214
Pevsner, Nikolaus, 238n29
Philadelphia Centennial Exhibition, 15, 73, *74*, 124, 128, 140, 141, 206
Philadelphia Gas Works, gasholders at, 205
Philebus (Plato), 218–219
Philipon, Charles, 56
Philosophical Inquiry into . . . the Sublime and Beautiful, A (Burke), 225n22
Philosophical Letters (Descartes), 81
photography, 21, 102–104, 227n41
Physiognomies des Chemins de Fer (Daumier), 57

Piazza d'Italia (Moore), *220*, 220–221, 249n41
Piazza San Marco clock tower, 82
Picabia, Francis, 215
Pickwick Papers (Dickens), Seymour's illustrations for, 45, 231n23
picturesque, the, 12, 86, 225n22. See also landscapes
"Pioneer" luxury railroad car, 233n46
Pissarro, Camille, 101
Pittsburgh, steel industry at, 226nn25–26
plating, 113. See also electroplate; electrotyping; silverplate
Plato, 218–219
Pleasures of the Rail-Road, The (Hughes), 4, 5
plumbing, 212
Pneumatica (Hero of Alexandria), 82, 187
pneumatic tube, 43
Pocock, George, 38
Podsnap family (Dickens), 137
"Poet, The" (Emerson), 14
political cartoons, American, 226n30
pollution: air, 8, 10, 42, 43 (*see also* smoke); social, 47
Pompeii excavations, 187
Pop Art, 215
Pope, Alexander, 8
Porter, Charles T., 74
postmodernism, 220–221, 249nn18–19
pottery, production, 105
Powerhouse Mechanic (Hine), 102, *103*, 104
Practical Cabinet Maker, Upholsterer, and Complete Decorator, The (Nicholson), 246n21
Practical Essay on the Strength of Cast Iron and Other Metals, A (Tredgold), 181
precision, 165, 245n13
Precisionism, 1, 212, 223n1
Price, Sir Uvedale, 225n22
Prince, John, 184, *185*
Principles of Decorative Design, The (Dresser), 112, 124–125, 175, 176
Priorslee Ironworks, 186
progress: egalitarian consumerism as, 140; through industrialization, 229n52

"Progress Vase" (Reed & Barton), *128*, 128–129
Proportions of the Human Body (da Vinci), 100–101
prosthetic devices, 87–88, 90
Prussian cast iron, 145, 147–149, 242 n4, 243 n18, 243 n22
Pry, Paul, *see* Heath, William
Puck (U.S. weekly), 226 n29
Pugin, Augustus Welby, 21, 109, 117, 136, 146, 155–156, 157, 159, 167
Pullman, George, 233 n46
Pullman cars, 54, 233 nn46–47
pumping engines, 205
Punch (magazine), 57, 64, 226 n29
Putnam machine tool works (Fitchburg, Mass.), 199–200, 206

railroads, 13, 30–32, 46–51, 217, 226 n28; accidents on, 3–5, 15, 17, 30, 32, 48–64 *passim*, 105, 192, 232 n33, 232 n49; in American political cartoons, 226 n30; art calms public fears of, 35, 53–54; chromolithographs of, 223 n7; class differences on, 59–60, 62–66; humorous depiction of, 54–63 (*see also* satiric artists); idealization of, 5, 53–55; illicit intimacies while riding, 65–66, 234 n51; passenger comfort, 64–66; station design, 152, 191–192, *192*, 246 n23. *See also individual railway companies*
Railway Journey, The (Schivelbush), 18
Rain, Steam, and Speed—The Great Western Railway (Turner), 34, 216
Rasl, Zdenke, 242 n4
Read, Herbert, 183
Réaumur, R.A.F. de, 110, 139
Redgrave, Richard, 120, 238 n30
Reed, Henry, 133
Reed & Barton, 109, *128*, 128–129, 133, 134, 240–241 n15
"Reflections on Technology" (Smith), 239 n5
Renaissance revival in electroplate, 134
Renwick, James, 33
repetition, 155
replicas: of artworks, in cast iron, 149, 151, 166; by electrotype, 112, 130, 134, 138; modern, of 19th-century cast iron, 242 n26

Return, The (Solomon), 65, 65–66
"Review of the Arts and Crafts" (Loos), 209–210
revivals, stylistic: Egyptian, 190, 245 n16; Gothic, 110, 135, 171, 190, 245 n15; Greek, 246 n18 (*see also* neoclassicism); Renaissance and rococo, 134
Reynolds, Edwin, 208
River Rouge factory, Ford Motor Company, 1
Robertson, George, *68*, 68–69
Robida, Albert, 88
Robinson, Eric, 109, 236 n9
Rocket, locomotive, 4, 31, 49
"Rocket" Oldsmobile, 50
rocket-powered travel, 49–50, *50*, *51*
rococo revival, 134
Rogers, William, 240 n14
Rogers Bros., 240 nn14–15, 241 n15
rolling mill engine, Hargreaves and Company's, 202
Rolling Power (Sheeler), 217, *218*
Romantic and Picturesque Scenery of England and Wales, 12
Rosenquist, James, 215
Rowlandson, Thomas, 36
Royal Academy, 150
Royal Dockyard, Portsmouth, 184
Royal Pavilion, Brighton, 152
runaway machines, 17, 30, 32, 34, 39, 46, 57, 58, 60, 179, 193
runaway technology, 70
Ruskin, John, 18, 20, 21, 100, 106–121 *passim*, 129, 136–137, 146, 156–157, 162, 165, 166, 211, 221, 228 n42, 246 n23
Russian Futurism, 216–217

Sacrifice of Isaac, The (Brunelleschi), 130
safety, 223 n7. *See also* accidents; explosions
Saint Charles Hotel, New Orleans, 153
St. John steamship explosion, 16, *16*
St. Petersburg (Russia) Academy of Sciences, 240 n5
Salford cotton mill, Boulton and Watt, 151
sand casting, 146, 147
"Sand-Man, The" (Hoffmann), 85

satiric artists, 4–5, 19, 31–56 *passim*, 67, 79, *80*, 88–94, 97–98, 215, 226n29, 230n15. *See also* Cham; Cruikshank, George; Daumier, Honoré; Heath, William; Seymour, Robert; Tenniel, Sir John
scale: human vs. machine, 7–15, 17, 68–79; shrinking world, 43
Schaefer, Herwin, 186, 245n13
Schamberg, Morton, 212, *213*
Scharf, Aaron, 227–228n41
Schinkel, Karl Friedrich, 149, 209
Schivelbush, Wolfgang, 18
Schlemmer, Oskar, 214
School of Art, Coalbrookdale, 120–121
Schools of Design, British, 26, 120, 123–124
Scientific American, 15, *76*, 76–77, 84–85, 87–88, 164, 232–233n44
screw-cutting machines, Whitworth's, 245n13
sculpture, cast-iron, 149–151
Second Class (Solomon), 233–234n50
second-class railway passengers, 232–233n44
Seegman, John, 228n50
Select Committee on art education, British Parliament, 120
Semper, Gottfried, 238n30
Senefelder, Aloys, 37
Seven Lamps of Architecture, The (Ruskin), 61, 106–107, 111, 129, 146, 157, 221, 246n23; "The Lamp of Truth," 136, 156
Severn River, Iron Bridge over, 224n16
sewing machines, 199, *199*, 200–201
Seymour, Horatio, 226n30
Seymour, Jane, 231n23
Seymour, Robert, 19, 36, 37, 42, 45–46, 79–97 *passim*, 214, 227n39, 231n23
"Shams and Imitations" (*Journal of Design and Manufactures* article), 22
Shaving by Steam (Seymour), 94
Shaving Machine sold by D. Merry and Sons, The (satiric print), *97*, 97–98
shearing frame, 11
Sheeler, Charles, 1, *2*, 212, 217, *218*, 223n1
sheet-iron houses, British and Belgian, 242n22
Sheffield, 108, 109, 131, 133, 136, 139

Sheffield plate, 105, 131
Sheffield Society of Artists, 106
Shelley, Mary, 69, 85–87
Sheridan, Richard, 12
Shingle Style architecture, 28
ship design, as analogue for machine, 203, 204
"Signalman, The" (Dickens), 48–49
signals, railway, 52
"Signs of the Times" (Carlyle), 79–81
Silliman, Benjamin, 130, 138, 139, 186
"Silver Palace" luxury railroad car, 233n46
silverplate, 108, 131–132, 237n9, 240nn14–15. *See also* electroplate; Sheffield plate
simplicity, 125, 144, 194, 207–208, 245n13
Simpson, Hall and Miller, 133
simulations, 228n43. *See also* automatons; computer-assisted imagery; imitations
Singer, Isaac, 199
Singer Sewing Machine Company, 199
sleeping cars, 64, 233nn46–47
Smee, Alfred, 130
Smith, Cyril Stanley, 109, 112–113, 119, 239n5
Smith, W. T. Russell, 226n26
Smith, Walter, 26, 116, 117, 121, 122, 128
Smith and Founder's Director, The (Cottingham), 190
smoke, 66, 106. *See also* pollution
Soane, Sir John, 189
Soho works, Boulton's, 105, 109, 131
Solomon, Abraham, *65*, 65–66, 233–234n50
Somerset House design school, 120
South Carolina Railroad, 34
Southern Belle, steam engine, 205
South Kensington Museum, 130, 151
South Kensington School of Design, 123–124
speed, 4, 30, 31, 38–60 *passim*, 216, 217, 248n10. *See also* runaway machines
speed limits, railroad, 52
Spencer, Thomas, 130, 239n5
spinning machine, 8, 224n10
Spiritalia (Hero of Alexandria), 187. *See also Pneumatica*

stability: architectural, 177; in images of speed, 217; and neoclassicism, 190, 191, 192–193; in steam-engine design, 179, 198
stagecoach travel, 52
stamping of silverplate, 240n15, 241n36, 241n40
standardized parts, 84, 109, 236n8
status: aesthetic criticism and, 22–23, 137; and critiques of imitative arts, 137; hallway furniture as symbol of, 172; of industrial designers, 107–108, 121–125; machine ornament and, 200; mill owner's, and neoclassical steam engine, 190; in tasteful consumption, 141–143
"Steam Arm, The" (Miller), 91–92
Steamboat on the Ohio (Anshutz), 35–36
steam carriage, 34, 38–43, 44, 231nn21–22; aerial, Henson's, 90
Steam Coach Passenger Set Down, A (Seymour), 46
Steam Engine, The (Tredgold), 181–182, *183*
steam engines, 178–181; Gothic revival, 245n15. *See also* boilers; explosions; locomotives
Steam Hammer at Work (Nasmyth), 71–72, *72*
Steam Man of the Prairies, The (Ellis), 83–84
steam-powered generator for electroplating, 132
steamships, 13, 30, 33
Steam Wagons in the Year 1942 in Vienna (engraving), 43
Steiner House, Vienna (Loos), 210
Stella, Joseph, 1, 217, 249n11
stencils, 199
Stephenson, George, 4, 31
Stern, Robert A. M., 220, 249n18
Stevens, Francis Simpson, 249n11
Stevens, Levi, 206
Stickley, Gustav, 113
Stockton and Darlington railroad line, 17, 46, 64
Storm King on the Hudson (Colman), 14, *15*
stoves, cast-iron, 147, *148*, 171
Streamlined Moderne, 217
Street Light (Balla), 211, 212

Stuart, James, 189
Stubbs, George, 118
style, revivals in: Egyptian, 190, 245n16; Gothic, 110, 135, 171, 190, 245n15; Greek, 246n18 (*see also* neoclassicism); Renaissance and rococo, 134
sublime, the, 12, 86, 225n22; technological, 226–227n31
sugar cane mill engine *Yankee Girl*, 187, *189*
Sullivan, Louis, 19, 177
Summerly, Felix, 25. *See also* Cole, Henry
Sutton Company beam engine, 186, *187*

Talbot, William, 21
"talking" doll, 84–85
"Tartarus of Maids, The" (Melville), 100
taste: and cast iron, 159–161, 167–168; elevation of, 138–144 *passim. See also* aesthetics; replicas
Tavernier, Jules, 32
Taylor, Herbert, 41
"Technical Instruction" (Morris), 23, 119–120
Technics and Civilization (Mumford), 182–183
technology, social responses to: acceptance, 13–15, 101, 102–104, 170–176; anxiety, 3–11 *passim*, 15–19, 31–34, 35, 70, 193, 200; artists' taming of, 35, 53–55, 65–66; automatons (*see* automatons; *Frankenstein*); framing through caricature, 93; integration, 20, 24–25, 29, 41, 108; in landscape painting, *9*, 9–15 *passim*, 35–36; legitimization, 7–8, 27, 34, 114–125, 142–144, 159–172 *passim*, 190, 246n23; in poetry, 7–8, 10, 12, 225n24; pride, 67; riot, 11, 42, 225n21; satiric art (*see* satiric artists)
Telephone (Schamberg), 212, *213*
Temps Nouveaux, Les (newspaper), 101
Tenniel, Sir John, 64, 233n49
textiles, 8, 113, 139
Thames tunnel banquet, 55
Theophilus, metallurgist, 131
Third-Class Carriage, The (Daumier), 63, *63*

266 Index

third-class railway passengers, French, 59–60, 63, *63*
Third Reich neoclassicism, 191
"Thirty Seconds at the Station" (Daumier), 61
Thompson, John W., 162
Thonet chair, 209–210
Thoreau, Henry David, 48
Thoughts on the Imitation of Greek Works (Winckelmann), 188
Tiger, steamship, 182, *183*, 245n6
Times (London), 30, 201–202
Tinguely, Jean, 216
Tolstoi, F. P., 240n5
Torso (Pevsner), 214
toymakers, 83, 84
trademarks, silverplate, 240–241nn14–15, 241n40
Train in the Countryside, The (Monet), 66
trauma, *see* accidents; anxiety; explosions
travel anxiety, 18, 47
travel regulations, railway company, 52
Travels, Observations, and Experiences of a Yankee Stonecutter, The (Greenough), 204
Treatise on the Sensations (Condillac), 86
Treatise on the Steam Engine (Renwick), 33
Tredgold, Thomas, 181–182, *183*
Trevithick, Richard, 30, 40, 180
triglyphs, 197
"Triumph of Science and the Industrial Arts" (Elkington, Mason, & Co.), 126, *127*
True Principles of Pointed or Christian Architecture, The (Pugin), 136, 155–156
Turner, J. M. W., 34, 49, 216
turnstile, 98–100
Two Paths, The (Ruskin), 118, 157
Tyson engine, 200

uniformity system, 84. *See also* American System
United States Centennial Exhibition, Philadelphia, 15, 73, *74*, 124, 128, 140, 141, 206
"Universal Infidelity in Principles of Design" (London *Times* article), 116, 238n29
University of Jena, 123
urn, neoclassical: as engine governor weight, 180; in stove design, *148*, 171
utilitarian design, 201–202. *See also* functionalism

Vaucanson, Jacques de, 82
Vechte, Antoine, 119
veneer, 113, 142
vernacular tradition, 245n13
Victoria and Albert Museum, 130
View in White Chapel Road, A (Aiken), 43
Villa Foscari (Palladio), 189
Villa Rotonda (Palladio), 189
Viollet-le-Duc, Eugène, 209
Voltaire, François Marie Arouet, 224n13
Vorticism, 249n11
Voss, Richard 248n3
vulgarity, 136, 137, 156, 167. *See also* class, social

Ware, Isaac, 195
Warhead (Burson), 215
Warhol, Andy, 215
Washington Iron Works (New York), 187, *189*
Watch (Murphy), 248n10
"water plant" bench, Dresser's, 175, *175*
waterpower, 225n25
Watt, James, 37, 40, 126, 178–181
Webster, Thomas, 190
Wedgwood, Josiah, 105, 107, 118, 236n8
Weimar Republic artists (Germany), 94, 214
Weir, John Ferguson, 77, 77–78, 79
Wellington, Arthur Wellesley, 1st Duke of, 4
Westinghouse, George, 17, 52
wet-plate process, photographic, 21
"wetopes," Moore's, 220
Wheeler and Wilson Sewing Machine Company, 200–201
whistle and signal systems, railroad, 52
Whitacre Pumping Station, Staffordshire, 205
White, John H., 233n47
Whitworth, Joseph, 245n13
"Who kissed Her in the Tunnel?" (*Leslie's* cover), 234n51
Wild Oats (U.S. weekly), 226n29
Williams, William, 9–10, 14, 224n16

Winckelmann, Johann Joachim, 188, 209
Wood, Robert, Philadelphia foundry of, 166–167
Wood & Perot, 149
Woodruff, Theodore, 233n46
woodworking machines, 110, 200
Wordsworth, William, 10, 224n19
"Work of Art in the Age of Mechanical Reproduction, The" (Benjamin), 227n41
"Work of Iron in Nature, Art, and Policy, The" (Ruskin), 156–157
Works in Iron (Matheson), 164–166, 169–170
World of Science, Art, and Industry, The (Silliman), 139
World War I, 214

Wornum, Ralph Nicholson, 138
Wreck of a Moody and Sankey Excursion Train (*Leslie's* illustration), *53*
Wright, Frank Lloyd, 28
Wright of Derby, Joseph, *78*, 78–79, 107, 118
wrought iron, 110, 156
Wyatt, John, 224n10
Wyatt, Matthew Digby, 25, 115, 159–161, 198

Yankee Girl (steam engine), 187, *189*
You Are Trapped on the Earth So You Will Explode (Holzer), 216
Young Carpenter's Assistant (Biddle), Haviland's edition, 161

zinc castings, 149